ANALYTICAL METHOD DEVELOPMENT AND METHOD VALIDATION FOR ASSAY AND RELATED COMPOUND OF PRAMIPEXOLE DRUG AND FINISH PRODUCT.

MR. MOTIRAM HUNA PATIL

Acknowledgement

I am highly indebted to "Almighty" enabling me to reach this destination. Words at my I am highly indebted to "Almighty" enabling me to reach this destination. Words at my command are simply inadequate to express my sincere thanks and profound gratitude to my supervisor **Dr. Manoj Dattatraya Rokade, Sitec Labs** for his invaluable guidance, liberal attitude and encouragement during my research work. In fact, to work with him was a pleasurable, enriching and memorable experience of my life and I take this opportunity to pay my sincere thanks and best regards to him.

I am thankful to my supervisor, **Dr. Ajit Godbole, M.D. of Anant Pharmaceuticals Ambernath Mumbai** for their continuous guidance, support and encouragement that made the compilation of my research work possible.

I am grateful to my Principal **Dr. (Mrs.) L. Natrajan**, Vice Principal **Prof. Vasant. G. Phadnis**, Head of Department **Prof. (Mrs.) Prema Narayan** and my colleagues **Dr. (Mrs.) Mala Bhaumik**, **Prof. (Mrs.) Jaya Bhojwani, Prof. Arun. P. Rajale, Prof. Naresh Bhatia, Prof. (Mrs.) Jyoti Sadhwani, Prof. Sanjay Lalwani, Prof. (Mrs.) Bhagyashri Pathai, Dr. (Mrs.) Radhamani Nair, Prof. Janardan Saini, Prof. S. P. Deshpande and Prof. Sanjay Lala** for their constant support and encouragement.

My sincere thanks also go to all labmates for their fondness and co-operation for providing a stimulating and fun environment which encouraged me to work hard during all the time of disappointments. The sleepless nights we worked together and for all the fun we have had in the last years. I also owe my thanks to those who have been a part of my life but I failed to mention.

I owe my special thanks to my wife **Mrs. Chhaya**, brother **Mr. C. H. Patil**, daughter **Mrs. Kavita**, son in law **Mr. Deepakkumar**, my son **Mr. Dhiraj and nephews Dr. Yogesh and Mr. Yogendra** who always stood by me with their moral and emotional support.

(Motiram H. Patil)

LIST OF TABLES

Table No.	Table Title	Page No.
1	Regulatory Guidance Documents	34
2	Summary of Characteristics of various Analytical Procedures	35
2.1	Properties of Inertsil Column	75
2.2	Selection of initial HPLC chromatographic conditions	76
2.3	List of industrial committees and regulatory agencies	80
3	Result of System suitability-Specificity(Assay)	88
4	Result of System suitability-Specificity(Assay)	88
5	Identification Study of assay	89
6	Linearity and Range study for Pramipexole (102.0– 306.0 µg/ml)	95
7	RSD for RT and area	96
8	Result of System suitability-Accuracy (Assay):	100
9	Result of System suitability-Accuracy (Assay):	100
10	Accuracy and recovery of Pramipexole:	101
11	Result of System suitability-Precision (Assay):	105
12	Result of System suitability-Precision (Assay):	105
13	Method Precision Results of Pramipexole:	106
14	Result of System suitability-Solution stability (Assay	110
15	Result of System suitability-Solution stability (Assay):	110
16	Sample solution stability of Pramipexole	111
17	Result of System suitability-Intermediate Precision (Assay	115
18	Result of System suitability-Intermediate Preision (Assay	115
19	Intermediate Precision Study	116
20	Cumulative Intermediate Precision Study	116
21	Result of System suitability-Robustness - I (Assay):	120
22	Result of System suitability-Robustness - I (Assay):	120

23	Robustness-I (Flow rate changed by –0.1ml) Assay of Pramipexole	121
24	Robustness-I (Flow rate changed by –0.1ml) cumulative %RSD	122
25	Result of System suitability- Robustness-II (Assay):	125

LIST OF TABLES

Table No.	Table Title	Page No.
26	Result of System suitability- Robustness-II (Assay):	125
27	Robustness-I (Flow rate changed by +0.1ml) Assay of Pramipexole	126
28	Robustness-I (Flow rate changed by +0.1ml) cumulative %RSD	126
29	Result of System suitability- Robustness-III (Assay):	129
30	Result of System suitability- Robustness-III (Assay):	129
31	Robustness-III (oven Temp. changed by +2) Assay of Pramipexole	130
32	Robustness-III (oven Temp. changed by +2) cumulative %RSD	130
33	Result of System suitability- Robustness-IV (Assay):	133
34	Result of System suitability- Robustness-IV (Assay):	133
35	Robustness-IV (oven Temp. changed by -2) cumulative %RSD	134
36	Robustness-IV (Temp. changed by -2) cumulative %RSD	134
37	Result of System suitability- Specificity (Related substances):	139
38	Result of System suitability- Specificity (Related substances):	140
39	Identification Study of Related substances	141
40	Summary for Signal to noise ration of peak obtained during LOD	146
41	Prediction of LOD and LOQ levels w.r.t. working concentration	148
42	Signal to noise ration of peak obtained during LOQ Study.	149
43	RSD of peak areas obtained during LOQ Study.	149
44	Results of Linearity of Pramipexole	153
45	Results of Linearity of Pramipexole Imp A	154
46	Results of Linearity of Pramipexole Imp D	155
47	Results of Linearity of Pramipexole Imp B :	156

48	Results of Linearity of Pramipexole Imp E	157
49	Result of System suitability-Accuracy (Related substances):	160
50	Result of System suitability- Accuracy (Related substances):	160

LIST OF TABLES

Table No.	Table Title	Page No.
51	% Recovery of Pramipexole known impurities Accuracy at LOQ Levels	161
52	% Recovery of Pramipexole known impurities Accuracy at LOQ Levels	162
53	Results of Accuracy study of Pramipexole Imp	163
54	Result of System suitability-Precision (Related substances):	168
55	Result of System suitability- Precision (Related substances):	168
56	Results of Precision Spiked of Pramipexole Imp A	169
57	Results of Precision Spiked of Pramipexole Imp B	170
58	Results of Precision Spiked of Pramipexole Imp D	170
59	Results of Precision Spiked of Pramipexole Imp E	171
60	Result of System suitability-Solution Stability (RS):	174
61	Result of System suitability- Solution Stability (RS):	174
62	Sample solution stability of Pramipexole Impurities	175
63	Result of System suitability- Intermediate Precision (RS):	179
64	Result of System suitability- Intermediate Precision (RS):	179
65	Results of Intermediate Precision Study	180
66	Results of Intermediate Precision Study	181
67	Result of System suitability-Robustness I (RS):	185
68	Result of System suitability- Robustness I (RS):	185
69	Robustness-I Results of % recovery of Pramipexole	186

Table No.	Table Title	Page No.
70	Robustness-I Results of cumulative RSD % recovery of Pramipexole	187
71	Result of System suitability- Robustness II (Related substances	188
72	Result of System suitability- Robustness II (Related substances):	188
73	Robustness-II Results of % recovery of Pramipexole	189
74	Robustness-II Results of cumulative % RSD of Pramipexole	190
75	Result of System suitability- Robustness III (Related substances):	191

LIST OF TABLES

Table No.	Table Title	Page No.
76	Result of System suitability- Robustness III (Related substances):	191
77	Robustness-III Results of % recovery of Pramipexole	192
78	Robustness-III Results of cumulative RSD % recovery of Pramipexole	193
79	Result of System suitability- Robustness IV (Related substances	194
80	Result of System suitability- Robustness IV (Related substances)	194
81	Robustness-IV Results of % recovery of Pramipexole	195
82	Robustness-IV Results of cumulative RSD % recovery of Pramipexole	196
83	Resolution System suitability for Assay:	198
84	System Suitability for Pramipexole	199
85	Solution stability For Pramipexole	200
86	The % relative difference for the content of Pramipexole	201
87	Resolution System suitability for Related Substances	203
88	System Suitability for Pramipexole Imp A :	204
89	System Suitability for Pramipexole Imp B :	204
90	System Suitability for Pramipexole	205
91	System Suitability for Pramipexole Imp E:	205

92	System Suitability for Pramipexole Imp D	206
93	Results of LOD and LOQ	207
94	% RSD for recovery of Impurities	209

LIST OF FIGURES

Figure No.	Title Page	Page No
1	James Parkinson	39
2	Parkinson's disease	41
3	The substantia nigra	41
4	Agilent 1100 HPLC Model	54
5	Schematic Diagram of HPLC	55
6	Refractive index detector	57
7	UV absorption detector	58
8	Multi-wavelength UV-Vis Absorption Detector	59
9	Multi-wavelength UV-Vis Absorption Detector Absorbance	59
10	Fluorescence detector	60
11	Schematic Diagram of Fluorescence detector	61
12	Fluorescence detector Absorbance	62
13	Evaporating light Scattering detector	63
14	Chiral detector	65
15	Conductivity detector	66
16	Mass spectrometer detector	67
17	Amperometric detector	68
18	Amperometric detector	68
19	Electochemical detectors	69
20	ICP Detectors	70
21	Hallow Cathod detectors	71
22	Agilent 1100 HPLC Model with tube size	72
23	Modern HPLC Used as Agilent 1260 Infinity	73
24	HPLC Support Particles	74
25	Particle Size Considerations	74
26	New Inertsil ODS columns	75

27	Different Sources of Impurities	81
28	Pramipexole UV spectrum in water	84
29	Pramipexole UV Spectrum in Methanol	85
30	Chromatogram of diluent in specificity of assay	89
31	Chromatogram of placebo in specificity of assay	90
32	Resolution chromatogram of Pramipexole and its related Impurities in specificity of assay	90
33	standard solution chromatogram in specificity of assay	91
34	Pramipexole Imp A Solution chromatogram in specificity of assay	91
35	Pramipexole Imp B Solution chromatogram in specificity of assay	92
36	Pramipexole Imp D Solution chromatogram in specificity of assay	92
37	Pramipexole Imp E Solution chromatogram in specificity of assay	93
38	Linearity curve of Pramipexole	96
39	Chromatogram of Blank in Linearity of Assay	97
40	Resolution chromatogram of Pramipexole and its related Impurities in Linearity of Assay	97
41	standard solution chromatogram of 50% in Linearity of Assay	98
42	standard solution chromatogram of 150% in Linearity of Assay	98
43	Chromatogram of diluent in Accuracy of Assay	102
44	Chromatogram of placebo in Accuracy of Assay	102
45	standard solution chromatogram in Accuracy of Assay	103
46	Pramipexole Accuracy 50% Solution chromatogram	103
47	Pramipexole Accuracy 100% Solution chromatogram	104
48	Pramipexole Accuracy 150% Solution chromatogram	104
49	Chromatogram of diluent in Precision of assay	107
50	Chromatogram of placebo in Precision of assay	107
51	Resolution chromatogram of Pramipexole and its related	108

	Impurities in Precision of assay	
52	standard solution chromatogram in Precision of assay	108
53	Sample Solution chromatogram in Precision of assay	109
54	Standard solution stability Curve	111
55	Chromatogram of diluent in Solution Stability of assay	112
56	Chromatogram of placebo in Solution Stability of assay	113
57	Resolution chromatogram of Pramipexole and its related Impurities in Solution Stability of assay	113
58	standard solution chromatogram in Solution Stability of assay	114
59	Sample 0th HR Solution chromatogram in Solution Stability of assay	114
60	Chromatogram of diluent in Intermediate Study of assay	117
61	Chromatogram of placebo in Intermediate Study of assay	117
62	Resolution chromatogram of Pramipexole and its related Impurities in Intermediate Study of assay	118
63	standard solution chromatogram in Intermediate Study of assay	118
64	Sample Solution chromatogram in Intermediate Study of assay	119
65	Chromatogram of diluent in Robustness Study I of assay	122
66	Chromatogram of placebo in Robustness Study I of assay	123
67	Standard solution chromatogram in Robustness Study I of assay	123
68	Sample Solution chromatogram in Robustness Study I of assay	124
69	Chromatogram of diluent in Robustness Study II of assay	127
70	Chromatogram of placebo in Robustness Study II of assay	127
71	Standard solution chromatogram in Robustness Study II of assay	128
72	Sample Solution chromatogram in Robustness Study II of assay	128
73	Chromatogram of diluent in Robustness Study III of assay	131
74	Chromatogram of placebo in Robustness Study III of assay	131
75	Standard solution chromatogram in Robustness Study III of assay	132
76	Sample Solution chromatogram in Robustness Study III of assay	132
77	Chromatogram of diluent in Robustness Study IV of assay	135

78	Chromatogram of placebo in Robustness Study IV of assay	135
79	Standard solution chromatogram in Robustness Study IV of assay	136
80	Sample Solution chromatogram in Robustness Study IV of assay	136
81	Chromatogram of diluent in specificity of RS	142
82	Chromatogram of placebo in specificity of RS	142
83	Resolution chromatogram of Pramipexole and its related Impurities in specificity of RS	143
84	standard solution chromatogram in specificity of RS	143
85	Pramipexole Imp A Solution chromatogram in specificity of RS	144
86	Pramipexole Imp B Solution chromatogram in specificity of RS	144
87	Pramipexole Imp D Solution chromatogram in specificity of RS	145
88	Pramipexole Imp E Solution chromatogram in specificity of RS	145
89	LOD solution of Pramipexole Imp chromatogram	147
90	LOQ solution of Pramipexole Imp chromatogram	150
91	Linearity Plot of Pramipexole for Linearity Study	153
92	Linearity Plot of Pramipexole Imp A	154
93	Linearity Plot of Pramipexole Imp D for Linearity Study	155
94	Linearity Plot of Pramipexole Imp B for Linearity Study	156
95	Linearity Plot of Pramipexole Imp E for Linearity Study	157
96	Chromatogram of diluent in Accuracy of RS	164
97	Chromatogram of placebo in Accuracy of RS	164
98	Resolution chromatogram of Pramipexole and its related Impurities in Accuracy of RS	165
99	standard solution chromatogram in Accuracy of RS	165
100	Accuracy 50% solution chromatogram in RS	166
101	Accuracy 100% solution chromatogram in RS	166
102	Accuracy 150% solution chromatogram in RS	167
103	Chromatogram of diluent in Precision of RS	171
104	Chromatogram of placebo in Precision of RS	172
105	Resolution chromatogram of Pramipexole and its related	172

	Impurities in Precision of RS	
106	standard solution chromatogram in Precision of RS	173
107	Sample solution chromatogram in Precision of RS	173
108	Chromatogram of diluent in Solution Stability of RS	176
109	Chromatogram of placebo in Solution Stability of RS	176
110	Resolution chromatogram of Pramipexole and its related Impurities in Solution Stability of RS	177
111	standard solution chromatogram in Solution Stability of RS	177
112	Sample solution chromatogram in Solution Stability of RS	178
113	Chromatogram of diluent in Intermediate Study of RS	182
114	Chromatogram of placebo in Intermediate Study of RS	182
115	Resolution chromatogram of Pramipexole and its related Impurities in Intermediate Study of RS	183
116	standard solution chromatogram in Intermediate Study of RS	183
117	Sample solution chromatogram in Intermediate Study of RS	184

Chapter 1
Introduction &
Review of literature

1.1 Introduction:

The growth of pharmaceutical industry in last three decades has been a pointer towards the requirement of quality medicines to alleviate diseases, ill health and suffering to animals and mankind. The simplest therapy of known organic molecules like aspirin, sulphadiazine and naturally occurring medicines from plant origin have been replaced now by complex organic molecules, which are analogues, derived from multiple synthetic routes. In the event of usage of such molecules the establishment of analytical profile of these molecules together with the desired medicinal activity like safety and efficacy has been the focus of attention of world-renowned scientists and pharmacist.

Drug substances are seldom administered alone, but rather as part of a formulation. Each particular pharmaceutical product is a formulation unique into itself. In addition to the active therapeutic ingredients, a pharmaceutical formulation also contains a number of non-therapeutic agents. These agents are generally referred to as pharmaceutical adjuncts, excipients or necessities, and it is through their use that a formulation achieves its unique composition and characteristic physical appearance.

There are many different forms into which a medicinal agent used for treatment of a disease. Most commonly known dosage forms are capsules, tablets, injections, suppositories, ointments, aerosols and many more including modern drug delivery systems like use of prodrug.

An Analytical chemist used in the many studies and used in this documentation. Drug analysis and assay play important role in the method development, actual manufacture of drug and use of drugs in routine analysis. Drug analysis shows the quality and its specifications which is used for full scale production.

1.2 Analytical Chemistry:

It is a stream in which the information is gain by the mean of method development and analysis performed one the instruments [1]. Analytical chemistry differs from the other branches of chemistry in both its scope and approach. While the other disciples are aimed at acquiring knowledge and creating theories in their respective fields, analytical chemistry develops method and tools necessary to acquire information about the chemical composition, its changes over time, its spatial arrangement and the structure of molecules and crystals. Scientists of chemical, physical, biological and engineers have collect the information sample constitute and structure. The correct choice and efficient use of modern analytical instruments requires an understanding of the fundamental principles of operation of these measuring devices [2]. The student –analyst when attains such an understanding that, reasonable choices be made among the several alternative means of solving an analytical problem; then student- analyst will be aware of the pitfalls that accompany physical measurements, and only then will the student be sufficiently attuned to the limitations in sensitivity, precision and accuracy of an instrumental measurement[3].

Analytical chemistry is the science, which deals with methods for detection of active and excipients, identification of them and quantification of chemical, biological and microbiological species in matrices of chemical, biological and environmental importance. A analytical method used for identifying the no of molecules atoms in sample. [4]

An analytical method is the way in which an analytical chemist obtains the required information. Pharma analysis is also defined as analytical techniques in chemistry through which the analysis was done on drugs both as API as well as finish product i.e. formulation. In pharmaceutical industry generally a bulk drug is referred to as API where pharmaceutical product is referred to as Drug Product or Finished product.

However in academic and pharmaceutical industry many other ways of analysis done like as bio-analytical chemistry, drug metabolism studies and biotechnology.

Drug analysis means identification, characterization and determination of drugs. Drug assay refers to determination of drugs in mixtures such as dosage forms and biologic fluids.

Drugs may be gases, liquids or solids. Prior to the formulation and manufacture of dosage forms, bulk drugs must be properly identified (qualitative analysis) and analyzed for drug content and the related impurities (quantitative analysis). Qualitative and quantitative determination of sample is also necessary once drugs are used in animals and humans during experiment development and treatment of patients.

1.3 Analytical Tests included for finish product Analysis:

Analysis of bulk drugs and pharmaceutical products generally carried out for following tests[5]:

- Description
- Identification
- Dissolution
- Content Uniformity
- Assay
- Chromatographic Purity or Related impurities
- Residual solvent analysis

1.3.1 Description Test:
It is used to describe of active and there related impurities in drug product is what it is stated.

1.3.2 Identification Test:
It is used to identify the active as well as impurities in the formulation. It includes tests such as melting point, color reactions, ultraviolet spectra, infra-red spectra and relative retention time by TLC or HPLC.

1.3.3 Dissolution test:
It is done for active dissolved in disso medium and no of percentage of active calculated. The major properties are also calculated by vivo tests on blood samples by bio-analytical methods. But in routine it is done by on in vitro models, which obviously have to be correlated with the in vivo data.

1.3.4 Content Uniformity:
Consistency of dosage units is important. All dosage should have uniformly percentage distributed contents, i.e. Analysis of particular batch shows similar percentage of the active component. This involves individual dosage form assay. This wanted in small amount active contents i.e. between a few µg per dose to 50 mg per dose.

1.3.5 Assay Test:
Assay is the estimation of potency of an active principle in a unit quantity of preparation. Ideally assays should not only be specific for chemical entity under examination, but also be stability determining.

1.3.6 Chromatographic Purity or Related impurities:
It indicates the homologues, analogues and byproducts from the analysis of synthesis or degradation. Chromatography has checking Analyte purity and is currently giving the fingerprint of a synthesis. Impurities and degraded products are checked from detection level to 150% of working concentration. That is active components should be checked up to 1000 folds excess of major compound. Because of this sometime minute amount of related substance gets overlap on the active. Which effects the in peak purity test and which will be solved by using photodiode array detection in liquid chromatography. The general specifications for related known impurities are based on its toxicology, whereas limit for unknown impurities are set in accordance to maximum daily dose.

1.4 Development of new analytical methods:

In the last century's second half, the technological developments of instrumental analysis were so wide and rapid that today the field of analytical chemistry has expanded towards "Computer Based Analytical Chemistry." Mainly focus on control of impurities and content of drug of product, as it is in small amounts effected at drug safety and efficacy. So the new Analytical methods for consistent quality are developed for checking the shelf life of the product.

It is also necessary to remember the choice of the method, as a wrong choice can lead not only to erroneous results but also a loss of valuable time and money. During the developmental stage of method it is very important for the analyst to know the final goal of the method. Before initiating analysis, the analyst should consider the following points are as follows:

- What kind of sample matrix is to be analyzed?
- What are the analytes to be determined and what are their expected concentration levels?
- Are there likely to be any interfering substances in the sample?
- Are there any specific regulatory requirements, such as action level or reporting limit to be met by the analysis?
- What analytical instrumentation and analytical skills are available?
- Is it to be screening procedure capable of detecting and identifying a number of compounds with similar physical and chemical properties?
- The accuracy and precision required
- What level of detection is required?
- How robust should the method be? Is it intended for the use by skilled analyst or by technical assistant?
- How expensive the method is?
- The time required for completing the analysis.

After answering the above questions any of the analytical technique like classical or instrumental techniques can be applied. Before developing any chromatographic method, one has to review the nature of sample, and goals of the separation defined. The sample related information that needs to be known prior to HPLC method development is given below [6]

- ✓ Number of components present.
- ✓ Chemical structures (functionality of components)
- ✓ Molecular weight of compounds.

- ✓ pKa values, melting point and boiling points of the compounds.
- ✓ UV spectra of compounds
- ✓ Concentration of compounds in sample
- ✓ Sample solubility
- ✓ Nature of sample – regular or special.

After the analysis of sample by suitable method, the data generated has to be processed and logical conclusions should be prepared.

1.4.1 Purpose of analysis and define problem :

In order to define the problem, the following questions need to be answered,

- ➤ The characterization and isolation of pure component.
- ➤ Sometimes only main component and impurities separation is sufficient or other time it is required that main component should be separated from degradants or impurities from each other.
- ➤ If quantitative analysis is looked for then quantitation up to what level is required that should be known.
- ➤ Method development should be known to be done on how many different types of sample matrices that is raw material, formulations, environmental sample. Will one or more HPLC procedures require?
- ➤ Chromatographer should consider how many samples at a time. Sometimes for many samples to run, one has to opt for shorter run time.

1.4.2 Information of sample:

More the gathered information about a sample, the better is the beginning for the method development.

- ➤ No. of compounds in sample present ?
- ➤ Conc. range of compounds in the sample preparation?
- ➤ The other properties like the chemical structure, molecular weight, pKa value, UV spectra, solution stability and solubility.

1.4.3 Sample preparation and detection:

In HPLC analysis sample preparation is an important throughout the analysis as it provide a reproducible and homogenous results by injecting the column through auto sampler. It does not affect the sample retention and resolution. The detector is selected based on the information of UV spectra.

1.4.4 Mode of separation:
In HPLC there are two types of samples that need separation; that is the regular and special samples. Regular samples are usually the mixtures of small molecules (<2000Da), while special samples are those, which require specialized methodology for separation. Regular samples are those that do not require special sample treatment and can be separated using a regular method. Special samples are usually large macromolecules, enantiomers and various inorganic ions that require special sample pretreatment and specialized type of chromatography for their separation. Ionic solutes can be generally defined as organic molecules that contain one or more functional groups capable of acidic or basic behavior in the pH range of 2 to 8.

1.4.5 Selection of Stationary phase / Chromatographic Column:
In HPLC analysis column is very important for separation of analytes and impurities. The requirement of stable, high performance column is good for rugged, reproducible HPLC method.

1.4.6 Selection of MP:
The requirements are :
- High purity to avoid introduction of peaks that may overlap with the analyte peaks.
- Ready availability at reasonable cost
- Low viscosity and reactivity to avoid chemical interaction with the analytes.
- Immiscibility with the stationary phase.
- Compatibility with the detector. Thus, for absorbance detection, the solvent should not absorb at wavelengths to be used. For refractive index detectors, the solvent refractive index must be significantly different from that of the solutes.
- Limited flammability and toxicity.

Acn and MeOH are popularly used organic solvents in HPLC. ACN is usually used for the MP. Acetonitrile and water combinally used because it has lower UV cut-off (185 to 210 nm) and low viscosity leading to higher plate numbers and lower column backpressures. However methanol which has a relatively lower UV cut-off (205nm) is a reasonable alternative.

1.4.7 Selection of Detector:

The detector is chosen depending upon some characteristic property of the analyte like UV absorbance, fluorescence, conductance, oxidation, reduction etc. Characteristics that are to be fulfilled by a detector to be used in HPLC determination are:

- ✓ High sensitivity, facilitating trace analysis
- ✓ Negligible baseline noise, to facilitate lower detection
- ✓ Large linear dynamic range
- ✓ Response independent of variations in operating parameters, such as, pressure, flow rates, temperature, etc.
- ✓ Response independent of MP composition
- ✓ Low dead volume
- ✓ Nondestructive to the sample
- ✓ Stable over long periods of operation
- ✓ Convenient and reliable to operate
- ✓ Inexpensive to purchase and operate
- ✓ Capable of providing information on the identity of the solute.

Multi component pharmaceutical preparations contain more than one active ingredient with variable concentrations. Detecting each component at their maximum absorbance, the purpose of using HPLC for their simultaneous determination is lost. Hence in such cases one has to explore the possibility of selecting the wavelength at which all the components are detected. This single wavelength is referred to as **"most suitable wavelength"**. While selecting the wavelength, the interest of minor component in the formulation or component with low extraction co-efficient needs special consideration. Variable wavelength detector or a photo diode array detector is used.

1.4.8 Optimization of selected parameters:

One has to optimize the selected parameters merely by changing each and every parameter. This will provide when, what and how much each parameter affect.

1.4.9 Anticipating problems:

What can give wrong results? Or what precautions one should take during the method application so that method can serve the intended purpose. Like if hardness of the tablets increases during the stability and if the method is with whole tablets then extraction of the active from the tablet may require more shaking or sonication time as time require to disintegrate the tablets may increase. If solution turns hazy after standing for long time may lead to wrong results so method design should take care of all the possible variables and manual errors.

1.4.10 Validation of the method:

Validation of method is an necessary for submission of developed product. The aim of validation of method is to show that it is suitable for outine analysis. As per the ICH guideline Q2 R2 for Method development and validation, the official methods need not to be validated for entire validation parameters but merely to be verified for their suitability and feasibility.

- Mainly validation parameters are as listed below :

1. Specificity
2. Accuracy
3. Linearity and range
4. LOD
5. LOQ
6. Precision
    - Repeatability
    - Intermediate Precision
7. Ruggedness

1.5 Method validation:

Validation is a basically required for quality and reliability of the results for analytical applications. The in house procedures are used for developments and applications. The method is used for samples analysed. The excipient other than analytes in pharma industry is constant results of batch is based on the results of procedures and tests. Required limits of the method are limit values, often based on stability, in the assay of an analyte and in Imp, which is take large safety factors into account.

1.5.1 Necessity of Validation:

The main reason for method validation is that the law requires it. It states:

Within the FDA[7], the CDER is take care of safe and effective drugs are available to the public.

The ICH guidance's were developed to harmonize the registration requirements of method validation and they do not necessarily cover all requirements that may be required in other parts of the world. The United States Pharmacopoeia (USP) [8][9] is published by a non-government organization whose publications of standards for the analysis of pharmaceuticals *(US Pharmacopoeia)*. General Chapter <1225> within the USP covers the requirements for validation of compendia methods and is generally included in any discussion of guidance documents relating to method validation.

Table 1: Regulatory Guidance Documents:

Agency	Title	Published
ICH	ICH-Q2A "Text on Validation of Analytical Procedure"	1994
	ICH-Q2B "Validation of Analytical Procedures: Methodology"	1995
CDER	Reviewer Guidance: Validation of Chromatographic Methods	1994
	Submitting Samples and Analytical Data for Method Validations	1987
	Analytical Procedures and Method Validation	2000
	Bio analytical Method Validation for Human Studies	1999
USP	<1225> Validation of Compendia Methods	2003

1.5.2 Validation requirements for Method Type:

The class of drugs are required for analysis are listed below and defined, with an table how they are determined. Availability of procedure are defines which test and methods are to be used and for which material. It is depend on which purpose for the procedure is used. A summation in table is applicable to various analytical procedures is **Table 2**.

Table 2: Characteristics Applicable to various Analytical Procedures:

Characteristic	Class A	Class B		Class C	Class D
		Quantitative tests	Limit tests		
Accuracy		✓		✓	✓
Precision		✓		✓	✓
Robustness	✓	✓	✓	✓	✓
Linearity and range		✓		✓	✓
Selectivity	✓	✓	✓	✓	✓
Limit of detection	✓		✓		
Limit of Quantification		✓			

Class of drugs for different types of methods:

Class A: It is used for identification in bulk stage or finish product stage form.
Class B: It is used for detection and quantification in finish product stage form.
Class C: It is used for assay calculation in active or finish product form.
Validation of methods is done by four common types [10][11][12]

- Identification Tests – is used for identification analyte in a sample. It is done by spectrum, Chromatographic behavior, chemical reactivity, etc. and with the comparisons to the reference standard
- Quantitative test for impurities' It is mainly used for the peak purity analysis in the sample.
- LOD and limit of quantification test for impurities
- Assay determines the percentage of analyte in the sample.

The assay is to determine the quantitatively analyte in the finish product. For drug product, same validation parameters are apply when same methods are used e.g. dissolution.

Method validation parameters are as below [13] [14]:

1. Specificity
2. Accuracy
3. Linearity and range
4. LOD
5. LOQ
6. Precision
    a. Repeatability
    b. Intermediate Precision
7. Ruggedness
8. Robustness

The detailed description is as follows:

1.5.3 Specificity / Selectivity:

It is to determine that the actives are free from any contamination or overlapping of any Imp. When the impurities are known and available in market then it is injected and compared with the retention time.

1.5.4 Accuracy:

The recovery is determination of active in the dosage form is determined by adding known amount of active in placebo to cover both above and below the normal levels (80%, 100% and 120 %) expected in samples. The peak area responses and assay of active in the three ranges were calculated and the accuracy of the results was compared to the actual percentage of drug added and the percentage of drug recovered. This is to know how accurately the active is retrievable from the sample matrix. It is carried out at three different levels. The amount recovered is calculated. Percent recovery at each level is calculated and statistical calculation gives you % recovery which should be within 98-102%. Which gives the clear indication that method is accurate?

1.5.5 Linearity and range:

This parameter checks the detector performance for the selected method. Series of dilutions are prepared and analyzed as per the method. The graph is plotted for Concentration vs. response. The detector response is said to be linear if Correlation coefficient of the graph when calculated to 0.9999 - 1.0.

1.5.6 Precision:

The precision is defined as the closeness between sample results. It is calculated by mean results and individual assay results by relative standard deviation.

1.5.7 Repeatability (within-laboratory variation):

It is same as precision but the same analysis is done by different analyst and on next day. It is done on same batch or same homogeneous material of same batch.

1.5.8 Reproducibility:

It is the precision of analysis which is carried on different conditions

1.5.9 Sensitivity:

It is determined by small variations in concentration and the same will be calculated by slope

1.5.10 Limit of detection:

The limit of detection of active and its related impurities is determined by serially diluting at lower concentration range, extended fairly close to the expected LOD. The concentrations were selected based on the logic that the method must be enough sensitive to at least quantify less than 0.5 times LOQ.

Duplicate injections of linear concentration of impurities are injected. The detection responses is calculated by formula i.e. DL = [3.3∗ SyX/Slope]. Based on the above calculation minimum detection limit are established.

1.5.11 Limit of quantitation (LOQ):

Duplicate injections of each Imp /active injected to 1- 150% of the specified concentrations (approx.) are made to establish lod and loq. The responses were calculatd from the calibration curve using the formula i.e. DL = [10∗ SyX/Slope]. Based on the above calculation minimum quantitation limit are established

1.5.12 Ruggedness:
This parameter means calculation of degree of reproducibility of the results observed in precision using different instrument, analysts, days, etc.

The ruggedness of Assay test method of the dosage form is carried out by two different analytical persons on two different instruments on different dates to know the %RSD of mean assay results between the same sample.

1.5.13 Robustness:
When method is deliberately deviated by small amount its capacity to remain unaffected and gives the normal results is shown by Robustness.

It is evaluated by checking following parameters:
1. Effect on changing column oven temperature by ± 2°C
2. Effect on changing wave length by ± 2nm
3. Effect of pH of the MP by ± 0.2
4. Effect on changing MP composition by 5.0 ml.
5. MP flow rate changed by ± 0.1ml/min.

Method validation is very tedious procedures but it is necessary for the new developed method since it is shows the quality of data generated and it is directly linked to the quality of procedure. Time constraints often do not allow for sufficient method validations.

1.6 Parkinson's disease (PD) is the most common neurodegenerative movement disorder, affecting almost 1% of the population above the age of 60 years.

1.6.1 Parkinson's disease is a progressive nervous system disorder that affects how the person moves, including how they speak and write. Symptoms develop gradually, and may start off with ever-so-slight tremors in one hand. People with Parkinson's disease also experience stiffness and find they cannot carry out movements as rapidly as before - this is called **bradykinesia**.

1.6.2 Parkinson's also affects the voice - a British mathematician believes he has created a cheap and easy to carry-out test using speech signal processing algorithms to accelerate the diagnosis of Parkinson's disease.

1.6.3 Parkinson's also affects sense of smell - The scientists said "Smelling tests in doctors' offices are suitable for detecting hyposmia, but so too are tests conducted in public places such as pedestrian zones."

1.6.4 History of Parkinson's disease:

Auguste François Chomel, a French pathologist, John Hunter, a Scottish surgeon, Hieronymus David Gaubius, a German physician and chemist, and Franciscus Sylvius, a Dutch chemist, physiologist and anatomist, all described Parkinson's-type signs and symptoms during the 17th and 18th centuries.

Fig. 1: James Parkinson

James Parkinson (1755-1824) - an English apothecary surgeon, political activist, paleontologist and geologist, wrote *An Essay on the Shaking Palsy* in 1817. James Parkinson systematically described six people with signs and symptoms of the disease we know today as Parkinson's.

Jean-Martin Charcot (1825-1893) - a French neurologist. His studies between 1868 and 1881 are described today by medical historians as a "landmark in the understanding of Parkinson's disease".

Frederic Lewy (1885-1950) - a prominent American neurologist is best known for the discovery of Lewy bodies, characteristic indicators of Dementia with Lewy Bodies and Parkinson's disease.

Konstantin Nikolaevitch Tretiakoff (1892-1958)- a Russian neuropathologist. While writing his thesis for his doctorate at L'Assistance Publique des Hopitaux de Paris, France, he described the degeneration of the substantia nigra in cases of Parkinson's-he was the first to link this anatomic structure with Parkinson's disease.

Rolf Hassler (1914-1984)- a German pathologist. Hassler made important discoveries in the treatment of Parkinson's disease. In a 1938 published paper, he wrote that autopsies of Parkinson's patients showed that the most affected part of the brain was the substantia nigra pars pallidus, which lost many neurons and had an abundant accumulation of Lewy bodies. He became a pioneer in surgery for tremor.

Arvid Carlsson (1923) - a Swedish scientist who was awarded the Nobel Prize in Physiology/Medicine in 2000 for his work on dopamine, Carlsson is best known for his work with dopamine and its effects in Parkinson's disease.

Kazimierz Funk (anglicized as Casimir Funk 1884 - 1967) - a Polish biochemist. Until the arrival of "levodopa", anticholinergics and surgery were the only available treatments for patients with Parkinson's.

1.6.5 Cause:

Parkinson's disease is caused by the progressive impairment or deterioration of neurons (nerve cells) in an area of the brain known as the substantia nigra. This communication coordinates smooth and balanced muscle movement. A lack of dopamine results in abnormal nerve functioning, causing a loss in the ability to control body movements.

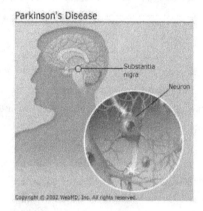

Fig. 2: Parkinson's disease is primarily caused by low and falling dopamine levels

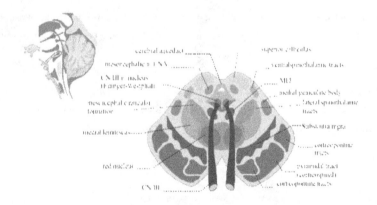

Fig. 3: The substantia nigra

1.6.6 Why Does Parkinson's Disease Occur?

Why Parkinson's disease occurs and how the neurons become impaired is not known. However, there is increasing evidence that Parkinson's disease may be inherited (passed on genetically from family members).

1.6.7 Some factors may raise or lower the risk of developing Parkinson's

Neurologist and co-author, co-author Brad Boeve, M.D., said:

> *"Understanding that certain patients are at greater risk for MCI or Parkinson's disease will allow for early intervention, which is vital in the case of such disorders that destroy brain cells. Although we are still searching for effective treatments, our best chance of success is to identify and treat these disorders early, before cell death."*

1.7 Pramipexole:

Pramipexole is a non-ergot dopamine agonist shown to be efficacious in the treatment of Parkinson's disease (PD). This review addresses the literature concerning pramipexole's efficacy in treating motor and non-motor symptoms in PD, its impact on the development of dyskinesias and response fluctuations, the issue of neuroprotection, and the risk for developing adverse events such as increased somnolence, attacks of sudden onset of sleep, cardiac valvulopathy and impulse control disturbances.

1.7.1 Pharmacodynamics:

Pramipexole is a non-ergotamine full agonist at the D2 subfamily of dopamine receptors, with higher selectivity for D3 than for D2 and D4 dopamine receptors[15][16][17].

1.7.2 Pharmacokinetics

Pramipexole has an absolute oral bioavailability greater than 90%, indicating good absorption and little first pass metabolism. Its elimination half-life is 8 hours in young, healthy volunteers[18][19][20].

1.7.3 Clinical efficacy

Summary of previous results: clinical trials enrolling 599 PD subjects[21][22][23][24]. On the Unified Parkinson's Disease Rating Scale (UPDRS)[25][26].

1.7.4 Adjuvant therapy in advanced Parkinson's disease

Summary of previous results: The combined results showed that pramipexole was well tolerated, associated with a reduction in daily L-dopa usage, and in daily "off" time, together with improvement in several parkinsonism and disability scorings[27][28][29][30]. The results of the second study were published recently [31].

1.8 Review of Literature:

Rotigotine: Responder rates and reduction in off-time were in absolute numbers somewhat higher for pramipexole, but not enough to show any statistically significance. Both pramipexole and rotigotine showed good tolerability, with more reports of hallucinations, dyskinesias and dizziness in the pramipexole group, and more reports of nausea and application site reactions in the rotigotine group[32].

Bromocriptine: Pramipexole was compared with bromocriptine and placebo as adjuvant therapy in 325 patients with advanced PD. Overall, pramipexole was shown to be clearly superior to placebo and not inferior to bromocriptine[33].

Guttman et al (1997) [34] compared adjuvant therapy with pramipexole and placebo in PD patients with a declining response to levodopa, including a bromocriptine group for comparison with placebo.

Rektorova et al (2003) [35] found no significant differences between pramipexole and pergolide as add-on therapies to levodopa in advanced PD patients with depression, as measured on UPDRS scales.

Hanna et al (2001) [36] In an open label study including 25 patients found no significant difference in the efficacy of pramipexole vs pergolide as adjuvant therapies in mild to moderate PD, although there was a slight trend in favour or pramipexole.

Goetz et al (2007) [37] investigated slow (over 8 weeks) vs rapid (over 1 day) switch from bromocriptine/pergolide to pramipexole in 16 PD patients on stable levodopa therapy.

Reichmann et al (2006) [38] conducted a study in 1202 PD patients who were switched slowly or quickly to pramipexole from any other oral DA, as add-on therapy, because of insufficient effectiveness on motor performance, tremor, and mood (depression, anhedonia).

Inzelberg R et al (2000) [39] reported that in an analysis of 7 trials on pramipexole, ropinirole, pergolide, entacapone and tolcapone as add-on therapies to levodopa in PD found that pramipexole and entacapone were the best choices for obtaining a greater reduction in levodopa dose and in "off" time, and a better tolerability.

Navan et al (2003) [40] conducted a randomized, double-blind, 3-month parallel study in 30 PD patients, most of whom were already medicated. The aim was to investigate the effects of pramipexole, pergolide, and placebo on parkinsonian tremor.

Pogarell et al (2002) [41] investigated 84 early and advanced PD patients with marked drug-resistant tremor under a stable and optimized antiparkinsonian medication. Pramipexole or placebo were administered as add on medication.

Pizzagalli et al (2007) [42] Cognitive impairment is common in PD, and up to a third of PD patients develop overt dementia.

Rektorova et al (2005) [43] examined in a randomized study the effects of pramipexole and pergolide as adjuvant therapy to levodopa in 41 non-demented patients with advanced PD and a current depressive episode.

Relja et al (2005) [44] evaluated over 6 months the effects of pramipexole as add-on therapy in PD patients treated with levodopa. An exclusion criterion was evidence of possible dementia and only patients scoring above 25 on the mini-mental state examination (MMSE) were included.

Brusa et al (2006) [45] investigated in a randomized study 20 patients with early/mild PD treated with levodopa or with pramipexole. Patients with dementia were excluded.

Biglan et al (2007) [46] found that better cognitive function defined as MMSE >28 was associated with a decreased risk for developing hallucinations after pramipexole therapy in PD.

Apathy: Apathy is a frequent finding in PD, significantly associated with cognitive impairment, executive dysfunction, anxiety and depression[47]. Apathy levels in PD are higher than in equally disabled patients with other diagnoses. While apathy does not seem to respond to levodopa, there are some indications from clinical trials that pramipexole's D3 agonist properties may be beneficial in this respect [48] [49].

Fantini et al (2003) [50] reported about 8 patients with idiopathic RBD, treated with pramipexole, 7 of which experienced a reduction in the frequency and intensity of RBD. In 5 patients this reduction was sustained.

Schmidt et al (2006) [51] reported about 10 patients with RBD, 3 of which had parkinsonism. Pramipexole reduced markedly the frequency and severity of RBD in 89% of the patients.

Fatigue: Fatigue is an early symptom in some PD patients [52]. While fatigue improved in patients with fibromyalgia with pramipexole treatment[53], it was reported as an adverse event in PD patients treated with pramipexole[54].

Dopaminergic complications in Parkinson's disease

The risk of experiencing response fluctuations (wearing off, on-off) or dyskinesias in PD has been evaluated at 40% after about 5 years of levodopa treatment[55].

After 2 years, 28 % of subjects assigned to pramipexole developed dopaminergic complications vs 51% in the levodopa group [56].

Ten patients in the pramipexole group developed dyskinesias prior to the open-label levodopa treatment, 7 of which had never been exposed to levodopa [57] [58] Initial pramipexole treatment was associated with the appearance of motor fluctuations before dyskinesias[59].

These findings are not unexpected, as dyskinesias seem to be mediated by the D1 receptor, for which pramipexole has a very low affinity compared with other compounds, such as cabergoline[60].

Neuroprotection

In vitro and in vivo studies

1-methyl-4-phenylpyridinium (MPP+) [61] [62], 6-hydroxydopamine (6-OHDA), rotenone[63], L-dihydroxyphenylalanine (L-DOPA), and methamphetamine may have a role to play also in the treatment of other neurodegenerative diseases related to beta-amyloid proteins[64], Pramipexole stimulated the production of a dopaminergic neurotrophic factor in tissue cultures[65], Finally, pramipexole reduced mitochondrial swelling and could thereby have an antiapoptotic effect[66]. In mice, pramipexole treatment completely antagonized the neurotoxic effects of MPTP [67] [68]. direct pharmacological effects of pramipexole on the ligand receptors, and not due to a slowing of the disease process with pramipexole [69]. Both animal and human studies show for example that short term therapy with pramipexole can down-regulate striatal dopamine transporters [70].

pramipexole delayed the point in time when levodopa treatment was needed and thereby delayed also the appearance of dyskinesias [71].

Using data from the CALM-PD study, **Noyes et al (2004)** [72] made a cost-effectiveness assessment of initial treatment with pramipexole vs levodopa, in early PD. Subjects with lower QOL and more depressive symptoms at baseline showed the highest improvement in QUALY gains over time[73] [74].

Paus et al (2003) [74] investigated 2592 PD patients, 284 of which were on a combination of pramipexole and levodopa therapy. Sudden sleep attacks were reported by 177 patients (6%) on phone interviews.

However, 25% of the patients have low scores on ESS[75] [76].

Etminan et al (2001) [77] analyzed in two separate analyses the risk for somnolence in PD patients taking pramipexole and ropinirole vs placebo (4 trials) and patients taking these two DA as adjuvant therapy to levodopa (7 trials).

Happe et al (2001) [78] found that sedation in PD may be a class effect of DA, with no difference between ergot and non-ergot DA.

O'Suilleabhain et al reported (2002) [79] on 368 PD patients treated with levodopa and DA, in monotherapy or in combination therapy.

Hobson et al (2002) [80] surveyed 638 PD patients, including 420 drivers. Patients taking different DA and levodopa, in monotherapy or in combination, were compared with respect to attacks of sudden onset of sleep.

Homann et al (2005) [81] identified 20 publications reporting sleep events in 124 PD patients, including 32 on pramipexole, 84 on other DA, and 8 on levodopa monotherapy.

Avorn J et al (2005) [82] from a study from on 929 patients, 39% of which used pramipexole, 18.5% ropinirole, 20% levodopa alone, and the rest a combination of different antiparkinsonian drugs.

Romigi et al (2005) [83] investigated one single PD patient with polysomnography. They found an increase in both the diurnal and the nocturnal sleep under pramipexole plus levodopa vs cabergoline plus levodopa therapy.

Razmy et al (2004) [84] conducted a polysomnography study in PD patients treated with pramipexole, ropinirole, bromocriptine or pergolide, finding that the total dopaminergic drug dose given, rather than the specific DA used, was the best predictor of daytime sleepiness.

Nieves AV et al (2002) [85] the available evidence shows that somnolence, while being part of PD itself, can be induced by all dopaminergic drugs, DA as well as levodopa itself. In the particular patient, switch to another DA may be attempted, as the susceptibility to particular DA may differ between patients. Modafinil treatment may be tried in difficult cases although the evidence supporting the use of modafinil in PD is conflicting [86] [87].

Peralta et al (2006) [88] investigated 75 PD patients and controls with echocardiography. Twenty-five of the patients were treated with pramipexole.

Zanettini et al (2007) [89] included in their study 36 PD patients treated with pramipexole, none of which showed any clinical important valvular regurgitation.

Junghanns et al (2007) [90] In a large British case-study involving 11417 people who were prescribed drugs for PD, there were 31 patients with cardiac valve regurgitation, but none of them had been treated with pramipexole[91].

Dewey et al (2007) [92] These results are not unexpected as the cause of valvular fibrosis appears to be fibroblast activation mediated through the 5-HT2B serotonergic receptors [93].

Chaudhuri et al (2004) [94] reported in a letter to the Editors of the journal *Movement Disorders* that they were "aware of pramipexole being implicated as well, at least in one case" in respect with "fibrotic reaction with non-ergot agonists".

Dodd et al (2005) [95] investigated 11 PD patients who had developed pathological gambling and other behavioral disorders.

Klos et al (2005) [96] reported on pathological hypersexuality in 13 PD patients, 6 of which were treated with pramipexole, and in 2 multiple-system atrophy (MSA) patients, 1 of whom was treated with pramipexole.

Weintraub et al (2006) [97] Shows the three patients who had the same DA treatment at follow up, one was taking pramipexole. His compulsive sexuality had gone into full remission after DBS surgery and decreased amantadine dose[98].

There is one case report of pathological gambling associated with pramipexole in a patient with restless legs syndrome[99].

This may lead to more reports on gambling and other behavioral disorders during pramipexole treatment, although the risk may be the same with other DA [100].

The development of hallucinations was one of the most common causes of study termination in patients with advanced PD (2.7% on pramipexole vs 0.4% on placebo) [101] [102].

The risk for developing hallucinations was higher in younger patients initially treated with pramipexole, compared with younger patients initially treated with levodopa, but this was not seen in the older cohort. Cognitive disturbances, older age and greater comorbid illness were associated with more hallucinations [103].

Tan and Ondo (2000) [104] reported on 15 PD patients and 2 RLS patients who developed pedal edema, out of 300 patients treated with pramipexole

Kleiner-Fisman et al (2007) [105] found a 7.7% (95% CI, 4.5%–12.9%) risk for development of pedal edema in the first year after initiation of pramipexole therapy in 237 PD patients, with more rapid development of edema among those with a history of coronary artery disease.

Introduction of HPLC and Impurities

Quality can be defined as the character, which defines the grade of excellence. The quality drug is something, which will meet the established product specifications, can be safely bought and confidently used for the purpose for which it is intended. There is no fear of adulteration or unpredictable side effects with such a quality drug.

It is important that analytical procedure proposed of a particular active ingredient or its dosage form should be systematically sound under the condition in which it is to be applied.

The test used for a drug quality assessment requires various analytical methods such as chemical, physicochemical, microbiological, biochemical and biological. The pharmaceutical quality control laboratory is composed of several divisions/sections for carrying out various tests requiring different analytical techniques.

1.9.1 Selection of suitable analytical methods are based on the following :

Nature of the sample to be analyzed.

Availability of the sample and concentration range which ensures sensitivity and range of concentration to be used for analysis.

Interfering components.

Physical and chemical properties of sample matrix.

1.10 Pharmaceutical Analytical Techniques:

Pharmaceutical analysis deals with the scientific and technical aspects of measurement of compositional and constitutional features of the sample.

It can be broadly divided into

1) Qualitative Analysis(Identification)
2) Quantitative Analysis(Estimation)

Qualitative Analysis reveals the identify of species i.e. the identification of compounds, elements or impurities in the sample.

Quantitative Analysis establishes the relative amount of one or more of these species or analyze in numerical terms. Qualitative information is required before a quantitative analysis is undertaken.

The various steps involved in a typical quantitative analysis are

- Chemical nature of the sample
- Obtain adequate amount of sample
- Selection of method
- Preparation of laboratory sample
- Number of samples to be analyzed
- Elimination of possible interferences
- Measurement of analyte
- Estimation of the reliability of the results(validation)
- The importance of newer analytical methods.

Drugs analysis means identification, characterization and determination of drugs. Newer analytical methods are developed for these drug combinations because of the following reasons:

1.11 Introduction to High-Performance Liquid Chromatography (HPLC)

The rapid growth of HPLC has been facilitated in the development of reliable, moderately priced instruments and efficient columns.

In classical column chromatography, the MP flows slowly through the column by means of gravity. But in HPLC, the separation is about 100 times faster than the conventional liquid chromatography due to packing of very small particles and thus differs from other Liquid chromatography technique. This small particle size results in more rapid approach to distribution equilibrium and smaller plate height(HETP) so that given length of column is highly efficient and peaks are narrow. However the close packing of these small particle reduces the flow rate of MP through the packed bed and in order to achieve a reasonable flow rate it is necessary to apply pressure to the MP.

This analytical method has prime importance for those compounds, which are volatile or thermally unstable so that they are not amenable to GLC analysis. The compounds which are not analyzed by GLC, but can be analyzed by HPLC are Carbohydrates, nucleoside, Steroids, alkaloids, peptides, amino acids and antibiotics. Prior to the development of HPLC, compounds are analyzed by GLC, Quantitative TLC, Paper chromatography and Liquid-Liquid or Liquid- solid chromatography.

1.11.1 Normal phase chromatography

Stationary phase : SiO_2, AL_2O_3, $-NH_2$, CN

Mobile phase : Heptane, hexane, Cyclohexane, dioxane, methanol.

1.11.2 Reverse-phase chromatography

Stationary phase : n-octadecyl (RP-18), n-octal(RP-8), ethyl (RP-2), phyenyl. $(CH_2)n$-CN, $(CH_2)n$-diol.

Mobile phase : Methanol or acetonitrile / water or buffer.

1.12 Instrumentations:

1.12.1 Agilent 1100 HPLC Model : (Figure. 4)

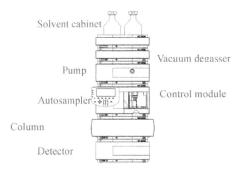

1.12.2 Schematic Diagram of HPLC: (Figure. 5)

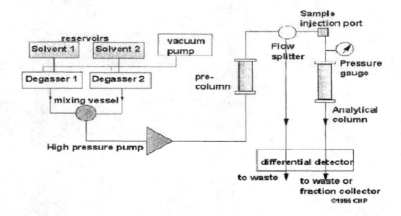

1.12.3. Type of Pumps in HPLC

1.12.3.1. Syringe Pump

1.12.3.2. Reciprocating pump
        Single piston reciprocating pump
        Dual piston
        Reciprocating diaphragm pump

1.12.3.3 Pneumatic Pump
        Direct pressure pump
        Amplifier pump

1.13 Injecting Systems

The Volume of sample used ranges from 0 to 500µl. The various injection methods are.

Syringe Injection
    Stop Flow Injection
    Loop Injection or Sampling Valve

1.14 Detectors

All detectors used in HPLC may be selective type and non selective type in which the former detects only a part of the components and the latter detects almost all the components.

The detectors used are

1.14.1 Refractive index detector : (Figure. 6)

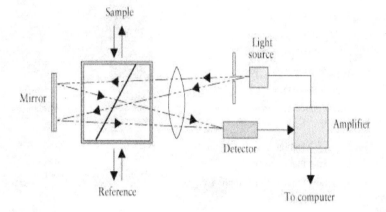

1.14.2 UV absorption detector :

Conventional UV-Vis Absorption Detector (**Figure. 7**)

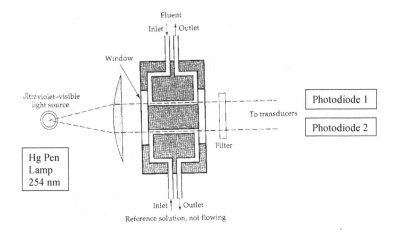

1.14.3 Multi-wavelength UV-Vis Absorption Detector: (Figure. 8)

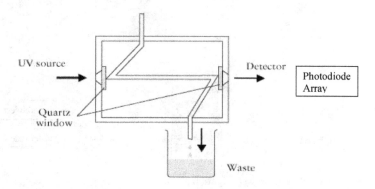

1.14.4 Multi-wavelength UV-Vis Absorption Detector (Figure. 9)

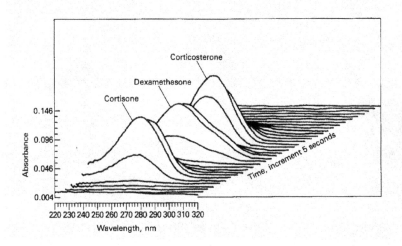

1.14.5 Fluorescence detector: (Figure. 10)

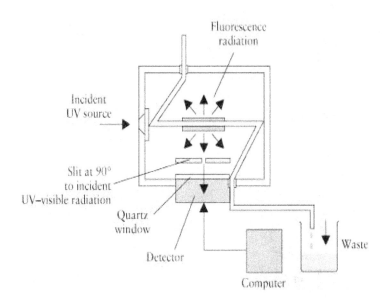

1.14.6 Schematic Diagram of Fluorescence detector : (Figure. 11)

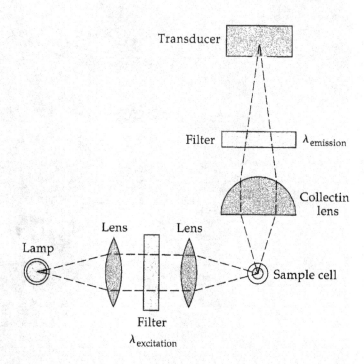

1.14.7 Fluorescence detector Absorbance : (Figure. 12)

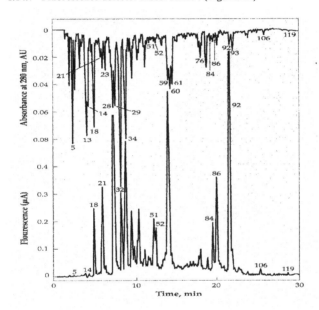

1.14.8 Evaporating light Scattering detector: (Figure. 13)

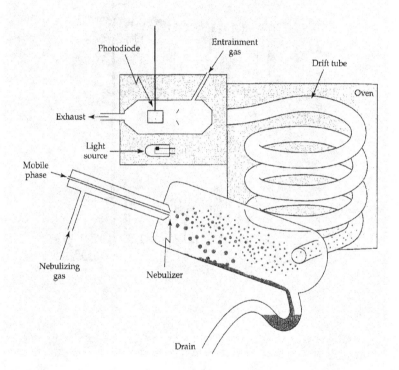

1.14.9 Evaporating light Scattering detector Advantages and Disadvantages:

Plus:

    1) Measures a bulk property

    2) Nearly Universal (must be non-volatile)

Minus:

    1) Signal not linear with concentration

    2) Fair sensitivity (LOD $\approx$ 1 ng)

    3) No salts or buffers in MP

1.14.10 Chiral detector : (Figure. 14)

1.14.11 Conductivity detector : (Figure. 15)

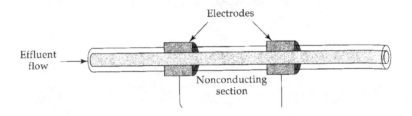

Plus:
    1) Common for Ion Exchange LC
    2) Detects species with no chromophores
    3) Simple, robust

Minus:
    1) Fair sensitivity (LOD ≈ 1 ng)
    2) No salts or buffers in MP
    3) Gradients a challenge

1.14.12 Mass spectrometer:

The Detector (**Figure. 16**)

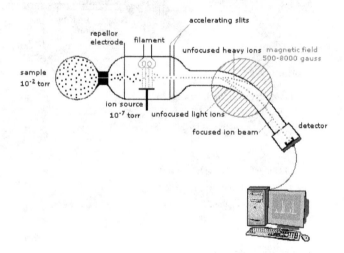

1.14.13 Amperometric detector : (Figure. 17)

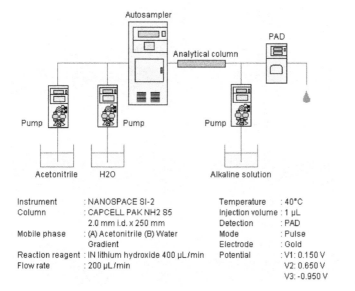

Instrument	: NANOSPACE SI-2	Temperature	: 40°C
Column	: CAPCELL PAK NH2 S5	Injection volume	: 1 µL
	2.0 mm i.d. x 250 mm	Detection	: PAD
Mobile phase	: (A) Acetonitrile (B) Water	Mode	: Pulse
	Gradient	Electrode	: Gold
Reaction reagent	: 1N lithium hydroxide 400 µL/min	Potential	: V1: 0.150 V
Flow rate	: 200 µL/min		V2: 0.650 V
			V3: -0.950 V

Selective detection of glucuronate conjugation in metabolite using pulsed amperometric detector

(Figure. 18)

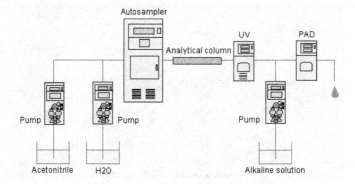

1.14.14 Electochemical detectors: (Figure. 19)

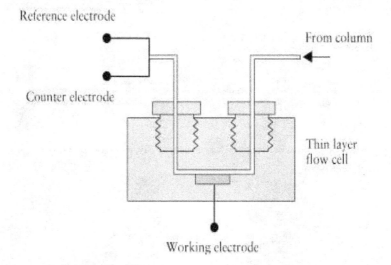

1.14.15 ICP Detectors: (Figure. 20)

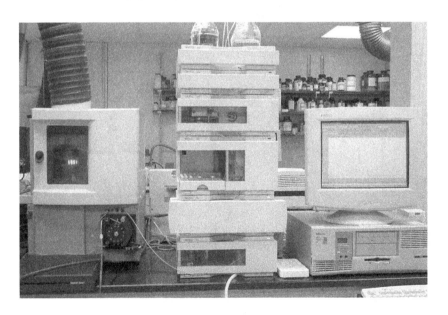

Plus:
    1) Can be Highly Selective
    2) Can be nearly Universal

3) Can have single sensitivity for all analytes

Minus:
    1) MP must be aqueous
    2) MP must not contain element of interest

1.14.16 Recent HPLC Detectors: (Figure. 21)

Hallow Cathod detectors:

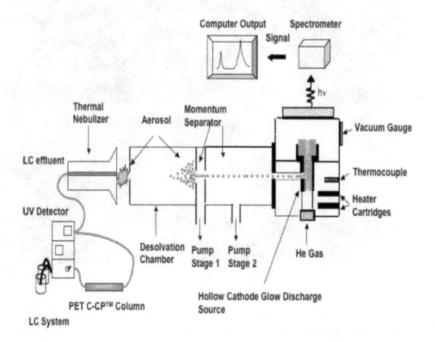

1.14.17 . Agilent 1100 HPLC Model (Figure. 22)

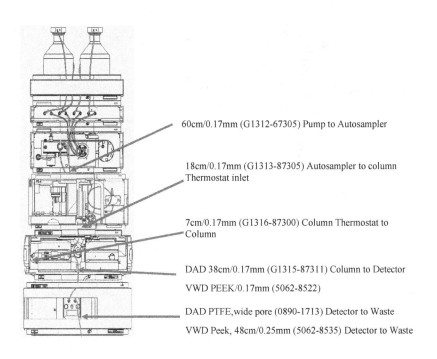

60cm/0.17mm (G1312-67305) Pump to Autosampler

18cm/0.17mm (G1313-87305) Autosampler to column Thermostat inlet

7cm/0.17mm (G1316-87300) Column Thermostat to Column

DAD 38cm/0.17mm (G1315-87311) Column to Detector

VWD PEEK/0.17mm (5062-8522)

DAD PTFE,wide pore (0890-1713) Detector to Waste

VWD Peek, 48cm/0.25mm (5062-8535) Detector to Waste

1.14.18 Modern HPLC Used as Agilent 1260 Infinity: (Figure. 23)

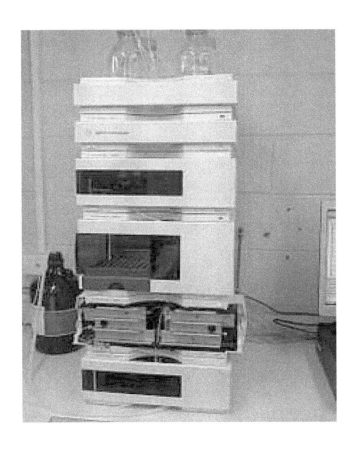

1.15.1 . **HPLC Column Used :**

HPLC Support Particles

Most often silica or silica gel (hydrated silicon-oxygen) (**Figure. 24**)

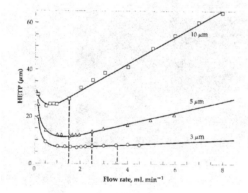

Less often hydrated aluminum-oxygen polymers (alumina)

Particle Size Considerations (**Figure. 25**)

1.15.2 New Inertsil ODS columns: (Figure. 26)

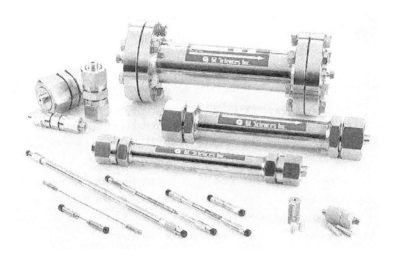

1.15.3 Properties of Inertsil Column : (Table 2.1)

Particle Size	Particle Shape	Surface Area	Pore Size	Pore Volume	Bonded Group	End-Capping	Carbon Load	pH Range
2μm	Spherical	450 m²/g	100Å	1.05 mL/g	Octyl Groups	Yes	9%	2 - 7.5
3μm								
5μm								
10μm								

1.16.1 ANALYTICAL METHOD DEVELOPMENT

HPLC method is used for quantitation or identification of drug the method should be a validated one. The method must be able to detect or quntitate the particular drug in presence of other components. Before starting the any method development one has to have knowledge about the information of the nature of sample, define separation goals.

The water soluble drug is further differentiated as ionic or nonionic which can be separated by reverse phase. Similarly, the organic soluble drugs can be classed as polar and non polar and equally separated by reverse phase. In some cases the non polar API may have to be separated using adsorption or normal phase HPLC, in which MP would be non polar organic solvent. The other chromatographic modes may need to be necessary for separation. These include ion exchange, chiral and size exclusion chromatography. In case of samples like proteins, peptides nucleic acids and synthetic polymers analyzed by using the some special columns or ion pair reagents (i.e. 0.1% TFA).

1.16.2 General conditions to initiate HPLC method development

Table – 2.2 Selection of initial HPLC chromatographic conditions

Chromatographic variables	Neutral compounds	Ionic-acidic compounds (carboxylic acids)	Ionic-basic compound (amines)
Column dimension (length, i.d.)	250 x 4.6 mm	250 x 4.6 mm	250 x 4.6 mm
Packing material	C_{18} or C_8	C_{18} or C_8	C_{18} or C_8
Particle size	5 μm	10μm or 5 μm	10μm or 5 μm
MP Solvents A and B	Water-acetonitrile	Buffer- acetonitrile	Buffer-acetonitrile
Ratio	50:50	20:80	20:80
Buffer and pH	No buffer required	Phosphate 25 M 3.0&7.5	Phosphate 25 M 3.0&7.5
Peak modifier	Do not use initially	1% v/v acetic acid	0.1% v/v triethylamine
Flow rate	1.0 to 2.0 ml/min	1.0 to 2.0 ml/min	1.0 to 2.0 ml/min
Temperature	Ambient	Ambient	Ambient
Injection volume	10μl to 25 μl	10μl to 25 μl	10μl to 25 μl
Sample concentration	< 100 μg	< 100 μg	< 100 μg

1.16.3 . Selection of MP

The selection of the MP mainly based on the solubility and polarity of the compound. Usually, in RP-HPLC method water and organic solvents are used as the MP. In NP-HPLC method non polar solvents like Hexane and THF were used.

In many cases, a silanophilic interaction causes tailing, mainly for the basic compounds MPs modifiers (0.1% v/v triethylamine for basic analyte or 1% v/v glacial acetic acid for the acidic analyte), or a combination thereof.

1.16.4 . MP composition

In reverse phase chromatography, the separation is mainly controlled by the hydrophobic interaction between drugs molecules and the alkyl chains on the column packing materials. This is due to the fact that a fairly large amount of selectivity can be achieved by choosing the qualitative and quantitative composition of aqueous and organic portions. Most widely used solvents in reverse phase chromatography are Methanol and Acetonitrile. Tetrahydrofuran is also used but to a lesser extent.

Initially experiments shall be conducted with MPs having buffers with different pH and solution organic phases to check for the best separation between the impurities. A drug solution having all possible known impurities can be used for checking the extent of separation with different MP ratios. Alternatively, solution of stressed drug substance can be used to check for separation of impurities. Silica based column with different cross linking's in the increasing order of polarity are as follows.

<-----Non-polar-----Moderately polar----Polar----->
C18<C8<C6<Phenyl<Amino<Cyano<Silica.

Experiments are to be conducted using different columns with different MP to achieve best separation in chromatography.

1.16.5 Selection of Buffer pH :

In reverse-phase chromatography analyte is nonpolar. It is undergoes ionization it becomes more hydrophilic and less interacting with column binding sites.

pH plays an important role in achieving the chromatographic separations by controlling the ionization. Drug molecules retention time depending on the pKa value Ex: Acids shows an increase in retention as the pH is reduced, While base a show a decrease.

1.16.6 . Selection of buffer

In reversed phase chromatograpy both condition (acidic and alkaline) potassium phosphate buffer pH 3.0 works well in general is an excellent buffer for analyte that contain acidic and amine functional groups. The potassium salt works better than the sodium salt for amines. It is important to use the buffers with suitable strength to cope up for the injection load on the column otherwise peak tailing may arise during chromatography. Therefore, strength of the buffer should be suitable enough to take injection load on the column so that peak tailing is avoided.

The retention times also depends on the strength of the buffers shall be between 0.05M to 0.20M. The selection of buffer and its strength is done always in combination with selection of organic phase composition in MP. But it is to be ensured that the higher buffer strengths shall not result in precipitants and turbidities either in MP or in standard and test solution while allowed to stand in bench top or in refrigerator. Experiments shall be conducted using different strengths to obtain the required separations. The buffers having a particular strength which gives separation of all individual impurities and degradants. The selected buffer strength and the effect of variation shall be studied.

1.16.7 . Selection of column

The HPLC column are cross linking the Si-OH groups with alkyl chains like, C8 (octylsilane), C18 (octadecyl silane) and nitrile groups (CN), phenyl groups (-C6H6) and amino groups (-NH2).

The following are the parameters of a chromatographic column which are to be considered while choosing a column for separation of impurities and degradants.

- Length and diameter of the column.
- Pore volume.
- Surface area.
- End capping.

1.16.8 . Selection of Column temperature

Always it is preferable to optimize the chromatographic conditions with column temperature as ambient. However, if the peak is symmetry could not be achieved by any combination of column and MP, then the column temperatures above ambient can be adopted. The increase in column temperature generally will result in reduction in peak asymmetry and peak retentions. When found necessary, the column temperatures between 30°C and 80°C shall be adopted. If a column temperature of above 80°C is found to be necessary, packing materials which can withstand to that temperature shall be chosen.

1.16.9 . Selection of flow rate

Flow rate shall be selected bases on the following data.
- Retention times.
- Column back pressures
- Separation of impurities.
- Peak symmetries.

Preferably the flow rate shall be not more than 2.5 ml/min. check the ruggedness of the method by varying by ± 0.2 ml from the selected flow rate. Select the flow rate which gives least retention tomes, good peak symmetries, least back pressures and better separation of impurities from each other and from API peak.

1.16.10 . Selection of Solvent delivery system

Chromatographic separation with a single eluent (isocratic elution) i.e.: All the constituents is mixed and pumped together as single eluent is always preferable. Gradient elution is a powerful tool in achieving separation between closely eluting compounds or compounds having widely differing in polarities. The important feature of the gradient elution which makes it a powerful tool is that the polarity and ionic strength of the MP can be changed (can be increased or decreased) during the run. Conduct experiments using different MP combinations and different gradient programme to achieve from API peak. In general, while running a gradient, two MPs having different composition is kept in different channels.

1.16.11 . Selection of detector wavelength

Selection of detector wavelength is finalization of the analytical method for impurities and degradants. Inject the Imp and API standard solutions into the chromatographic system with photodiode array detector and collect the spectra. Also conduct forced degradation studies and collect the UV spectra of all the major degradation products. Overlay the spectra of all the compounds and select a wavelength which is most common and gives higher responses for all compounds.

1.17 Method validation.

Table 2.3 List of agencies

Committees and regulatory agencies	Guidelines available
ICH	a) Q2R1Guidelines are guidelines b) Q1R1 Guidelines are for development and validation of stability indicating analytical methods includes methodology
The USFDA	Guidelines: a) for the validation of analytical methods2 b) For the validation of bio-analytical methods 3.
IUPAC[106]	"Harmonized Guidelines for Single-Laboratory Validation of Methods of Analysis".
EURACHEM[107]	detailed guide for method validation
AOAC[108]	verification of analytical methods for the ISO 17025 accreditation.
Huber [109]	documentation for the verification of Analytical methods for the ISO 17025 accreditation.
Viswanathan and co-authors [110]	An overview for validation of bio-analytical methods.

1.18 *Imp in pharmaceuticals:*

1.18.1 Different Sources of Impurities:

According to the ICH guidelines, impurities related to API's are classified in to the following main categories [111] to [131]

A. Organic impurities:

B. Inorganic impurities:

1.19. Residual solvents:

Figure No. 27 Different Sources of Impurities

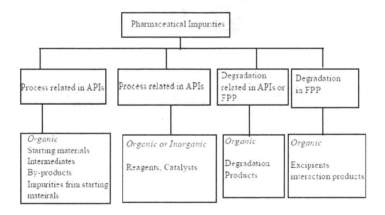

Chapter 2
Assay
Method Validation

2.1 Experimental Details and Chromatographic condition for Assay:

2.1.1 Instrument:

Chromatography was performed using HPLC of Waters 2695 Alliance system, Waters 996 with PDA detector. Chromatograms and data were recorded by means of Empower software version.

2.1.2 Chemical and Reagents:

United States Pharmacopeia reference standard of Pramipexole, Pramipexole A, Pramipexole B, Pramipexole D and Pramipexole E were supplied by Cipla Ltd , Mumbai .Samples containing Pramipexole was taken from Local market. Analytical reagent (AR) grade Potassium phosphate and Octane 1-sulphone sodium salt was purchased from Fluka (Banglore, India) and Acetonitrile from sigma Aldrich (Mumbai, India). Water for HPLC studies was obtained from Millipore water purifying system. GF/C filter paper was obtained from company name Whatmann. All preparation of standard and sample solution dilutions were prepared in class A volumetric flasks.

2.2 Method development for Assay and Related Substances:

2.2.1 Optimisation of MP:

The analytical method developed for detect and quantify the Pramipexole and there impurities by utilizing same chromatographic setup in single run eluted at reasonable time and proper separation of all the components. Optimization of conditions for simple, accurate and reproducible analysis involves analyzing system suitability solution on varying stationary phase, strength of aqueous phase, pH, different proportion water with acetonitrile, change in flow rate and at different column oven temperature.

Initially Reverse phase chromatography with MP containing buffers such as Potassium dihydrogen orthophosphate, Ammonium dihydrogen phosphate with organic modifier acetonitrile was tested in isocratic mode by varying the composition. The retention time of Pramipexole was found closely eluted which was not suitable due to the possible placebo interference .Also different concentrations of acetonitrile and different pHs did not make significant change in the elution pattern of Pramipexole and there impurities

Chemical structure of Pramipexole, Pramipexole, Pramipexole A, Pramipexole B, Pramipexole D and Pramipexole E .

MP A Contains Buffer prepared as by dissolved 4.5 gm of potassium phosphate and 2.0 gm of Octane sulphonate in to 2000 ml water , pH to 3.0 with diluted OPA): and Solution B is

Acetonitrile as 70:30 v/v. the flow is 1.0 ml/min and at 40°C. solution injected 5µl and 254 nm.

It is based on the peak shapes and resolution of Pramipexole, Pramipexole A, Pramipexole B, Pramipexole D and Pramipexole E and for resolution chromatogram refer fig. 32.

2.2.2 Selection of Chromatographic Column:

The polarity of Pramipexole compound are not strong so it is get retain on C_{18} column and not on C_8 column even if the MP has very low organic solvent, whereas Pramipexole and their related impurities shows good retention on a peerless Basic AQ C18 (250 x 4.6, 5 µm) column. Newer columns with minimize the potential problem of surface acidic, while separating basic compounds. So, C18, a porous, spherical packing material column was utilized as separation unit.

Wavelength scanning in 200nm to 400nm. Pramipexole show reasonably good response at 254 nm which is in line with Pharmacopeial wavelength for individual drugs. This is shown in **Figure 28**. A typical UV spectrum showing the scan of Pramipexole in water and methanol given in **Figure 29**.

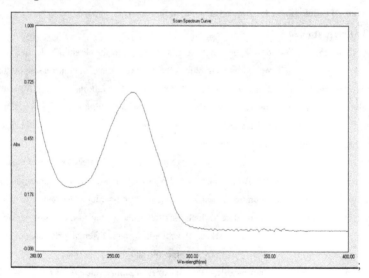

Fig. 28: Pramipexole UV spectrum in water

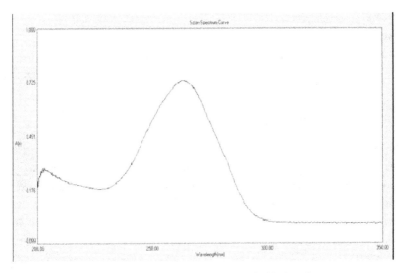

Fig. 29: Pramipexole UV spectrum in Methanol

2.3 Method Validation (Assay) :

2.3.1 Reagent and Reference Standards:

Pramipexole

Pramipexole Imp A

Pramipexole Imp B

Pramipexole Imp D

Pramipexole Imp E

Pramipex Tablets

Water (Milli Q).

Ammonium Acetate (GR Grade) - Merck

Acetonitrile (HPLC Grade). - Merck

2.3.2 Instruments and Equipments:

HPLC: Waters

Autosampler : Waters 2695

Detector   : Waters 996 (Photodiode Array Detector)

Pump   : Waters 2695

Software   : Empower 2

Version no : 6.20.00.00
Balance: Mettler Toledo.
Filters: Whatmann -GF/C filters – Cat no 1822-090 (0.45µm)
Whatmann- PVDF filter membrane GD/X (Poly- vinyl difluoride) – Cat no 6872-2504 (0.45 µm)
Whatmann filter paper Ashless 41 –Cat no 1441090 (20µm)

2.3.3 Chromatographic conditions:
HPLC column a peerless Basic AQ C18 (250 x 4.6 , 5µm) , with 1.0 ml/min and column oven temperature as 40° C. The buffer prepared by dissolved 4.5 gm of potassium phosphate and 2.0 gm of Octane sulphonate in to 2000 ml of water and pH to 3.0 with diluted OPA, filtered. Acetonitrile is the organic solvent used as MP B in isocratic mode. The injection volume amounted to 5µl. The analysis was carried out under isocratic condition as MP A: MP B is (70:30). at 254nm .Diluent used was a mixture of acetonitrile and water (50:50) The analytical method was analysed according to the ICH guidelines[132][133][134].

2.4 Specificity and System Suitability:
2.4.1 Specificity:
The methods validation parameter specificity is determined by comparing chromatograms of diluent, actives, the related impurities and the placebo of the

2.4.2 Preparation of Resolution solution for Assay:
Weigh accurately about 200 mg Pramipexole std, 1.5 mg of each Pramipexole Imp A, Pramipexole Imp B, Pramipexole Imp D and Pramipexole Imp E taken into 100 ml flask ,added 70 ml of diluent. Sonicate to dissolved and diluted up to the mark with diluent. Further diluted 10 ml to 100 ml with diluent. Filter with 0.45 µm syringe filter. The solution is injected and recorded the chromatograms and peak areas.

2.4.3 Preparation of Standard solution:
Weigh accurately about 200mg Pramipexole std into 100 ml vol. flask and added 70 ml of diluent. Sonicated for 5 minutes and dluted upto the mark with diluent. Further diluted 10 ml to 100 ml with diluent. Filter with 0.45 µm syringe filter. The standard solution prepared twice and injected for checking the similarity factor. Then the standard solution A was

injected 5 times separately. The solution is injected and recorded the chromatograms and peak areas.

2.4.4 Preparation of Sample solution:

Twenty tablets were weighed and ground to homogenous powder using a mortar and pestle. An accurately weighed portion of the powder, equivalent to 20 mg of Pramipexole was transferred into a 100 ml volumetric flask containing 70 ml of diluent and disperses with the aid of ultrasound for 10 minutes with intermittent swirling. The flask was further shaken with the means of mechanical shaker for 15 minutes and allowed to reach the ambient room temperature. The volume was made up to 100 ml with diluent and mixed. Filter through 0.45 µm filter paper using syringe. The solution is injected and recorded the chromatograms and peak areas.

2.4.5 Placebo Preparation:

An accurately weighed portion of the placebo, equivalent to 20 mg of Pramipexole was transferred into a 100 ml volumetric flask containing 70 ml of diluent and disperses with the aid of ultrasound for 10 minutes with intermittent swirling. The flask was further shaken with the means of mechanical shaker for 15 minutes and allowed to reach the ambient room temperature. The volume was made up to 100 ml with diluent and mixed. Filtered the solution through GF/C.

Calculation of Pramipexole:

$$= \frac{A}{B} \times \frac{W1}{100} \times \frac{10}{100} \times \frac{100}{W2} \times \frac{P}{100} \times \frac{100}{L.A.} \times Avg.\ wt$$

Where,
- A = Area of Pramipexole in sample solution
- B = Average peak area of peak in replicate injections of standard solution (A).
- W1 = Weigh of Pramipexole WS in Standard solution (A) in mg.
- W2 = Weigh of sample in gm.
- P = Purity of working standard
- L.A. = label amount of Pramipexole in Tablet
- Avg. wt = Average weight of Pramipexole tablets

System suitability was performed by injecting resolution solution and determining resolution between closely eluting peak of Pramipexole, Pramipexole A, Pramipexole B, Pramipexole D and Pramipexole E. The TF and TP were checked in Standard solution. The standard solution for Pramipexole was prepared twice and injected. The parameters such as similarity factor, and RSD of RT , Area of Pramipexole was determined. Results for resolution system suitability are presented in **Table 3 and Table 4**.

Table 3: Result of System suitability-Specificity(Assay):

Analyte	Resolution
Pramipexole Imp A	N.A.
Pramipexole Imp B	2.1
Pramipexole	2.3
Pramipexole Imp E	6.7
Pramipexole Imp D	4.6

Table 4: Result of System suitability-Specificity(Assay):

Analyte	% RSD		T.P.	T.F.	Sim. Fac
	RT	Area			
Pramipexole	0.0	0.4	12385	1.2	1.00

The acceptance criterion for similarity factor was ≥0.98 ≤1.02, tailing factor for the peak due to Pramipexole was NMT 2.0, theoretical plates was not less than 2000 and % RSD of RT , Area of Pramipexole Standard solution of first replicate solution was not more than 1.0% and 2.0% respectively. The acceptance criteria were met.

Diluent (**Figure 30**), Placebo Preparation **(Figure 31)**, Resolution solution and standard Preparation (**Figure 32**), were injected into the chromatograph. Peaks was not detected at the RT of Pramipexole and their respective impurities hence proving the specificity of the method. **Table 5** indicate the relative retention times (RRT) of all the impurities.

Table 5: Identification Study of assay

Sr. No	Components	Retention Time (mins)	Relative retention time (RRT)	Placebo peak area response
1	Pramipexole Imp A	3.062	0.69	Not detected
2	Pramipexole Imp B	4.007	0.91	Not detected
3	Pramipexole	4.407	1.00	Not detected
4	Pramipexole Imp E	5.824	1.32	Not detected
5	Pramipexole Imp D	6.847	1.55	Not detected

There were no peaks due to the MP, Diluent and Placebo at the retention time of Pramipexole and their impurities. This shows that method is specific by HPLC determination.

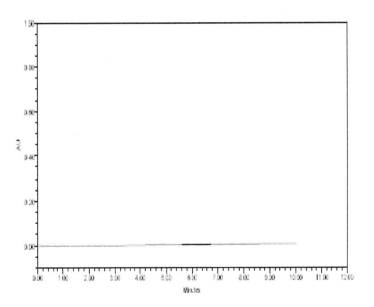

Figure 30: Chromatogram of diluent

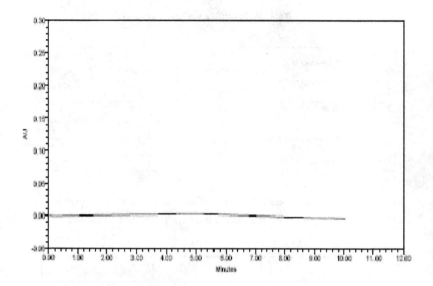

Figure 31: Chromatogram of placebo

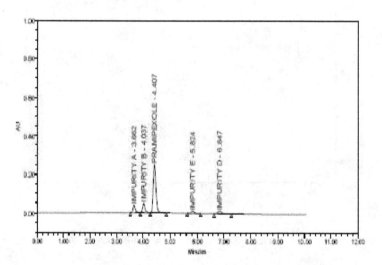

Figure 32 : Resolution chromatogram of Pramipexole and its related Impurities.

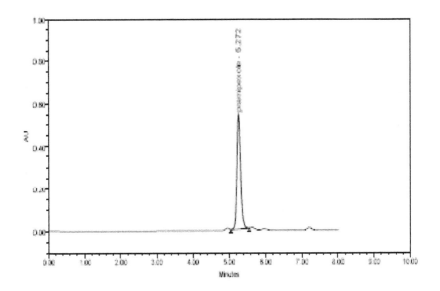

Figure 33: standard solution chromatogram

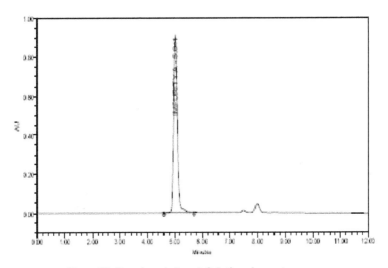

Figure 34: Pramipexole Imp A Solution chromatogram

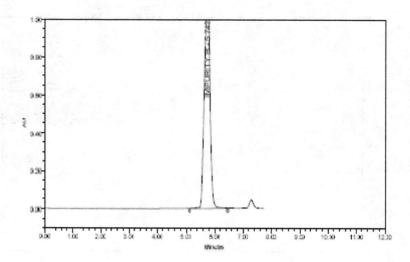

Figure 35: Pramipexole Imp B Solution chromatogram

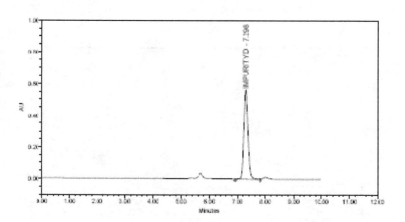

Figure 36: Pramipexole Imp D Solution chromatogram

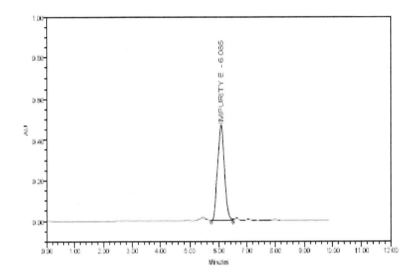

Figure 37: Pramipexole Imp E Solution chromatogram

2.5 Linearity and Range:

The concentration of serial standard solution is 50, 75, 100, 125 & 150% of assay concentration (working level 200 ppm for Pramipexole) was prepared and injected.

2.5.1 Linearity solution preparation:

Linearity Stock solution preparation:

Weighed 200 mg of Pramipexole WS into 100 ml of volumetric and mixed with 75 ml diluent. Shook well and diluted with diluent.

Preparation of 50% solution:

Pipetted out 5ml above linearity stock solution into 100 ml vol. flask and diluted up to mark with diluent. Then the solution was injected 6 times separately. The solution is injected and recorded the chromatograms and peak areas.

Preparation of 75% solution:
Pipetted out 7.5 ml above linearity stock solution into 100 ml vol. flask and diluted up to mark with diluent. Then the solution was injected 2 times separately. The solution is injected and recorded the chromatograms and peak areas.

Preparation of 100% solution:
Pipetted out 10 ml above linearity stock solution into 100 ml vol. flask and diluted up to mark with diluent. Then the solution was injected 2 times separately. The chromatograms are recorded and measure the peak response

Preparation of 125% solution:
Pipetted out 12.5 ml above linearity stock solution into 100 ml vol. flask and diluted up to mark with diluent. Then the solution was injected 2 times separately. The chromatograms are recorded and measure the peak response

Preparation of 150% solution:
Pipetted out 15ml above linearity stock solution into 100 ml vol. flask and diluted up to mark with diluent. Then the solution was injected 6 times separately. The solution is injected and recorded the chromatograms and peak areas.

2.5.2 Procedure:
First level (low conc. 50% solution) and Last level (high conc. 150% solution) is injected separately 6 times and other levels (exception of low & high conc.) are injected separately 2 times (5 µl).

The chromatogram is to be recorded and measured the peak response.

Calculation:
% Y-Intercept:

$$\% \text{ Y-Intercept} = \frac{\text{Y-intercept}}{\text{Mean area of working level conc.}} \times 100$$

Response Factor:

$$\text{Response Factor} = \frac{\text{Mean Response of each Linearity}}{\text{Concentration of Respective linearity level}}$$

A series of Solutions of standard Pramipexole was prepared over the range of 50 % level to 150% level of the specified limit for Assay. The results obtained are presented in **Table 6**.

Table 6: Linearity and Range study for Pramipexole (102.0– 306.0 µg/ml)

Sr. No.	Concentration Level	Concentration (µg/ml)	Mean Peak Area Counts
1	50%	102	1980782
2	75%	153	2971173
3	100%	204	3961564
4	125%	255	4951955
5	150%	306	5942346
Correlation Coefficient			1.000

Five Solutions containing concentrations of Pramipexole i.e. 102 µg/ml to 306 µg/ml. The r^2 for Pramipexole was determined as 1.000 , the Y-intercept for Pramipexole was -0.13%, the Response factor for Pramipexole 1.4% .

The criteria for regression coefficient was NLT 0.999, % Y-intercept was NMT ± 2.0% The % RSD of responses factor of Pramipexole peak was not more than 2.00%. The regression coefficient, the %Y-intercept, RSD of area and RT.

2.5.3 The Linearity co-efficient of Pramipexole:

The linearity curve is plotted between y-axis in mean response of injection replicate (peak area of Pramipexole) and x-axis in respective serial concentration (in ppm). The Linear co-efficient (R2) is 1.000 and the % y-intercept are to be determined and it is found -0.13% and .

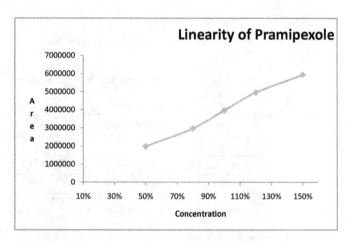

Figure 38: Linearity curve of Pramipexole:

2.5.4 Range:

The % of RSD for peak area and RT of Pramipexole from lower concentration (Lower level) and higher concentration (Higher level).

Table-7: % of RSD for Retention Time and peak area:

Name of the Standard	50 % level		150 % level	
	Retention time	Peak area	Retention time	Peak area
Pramipexole	0.3	0.8	0.2	0.2
Limit	NMT 1.0%	NMT 2.0%	NMT 1.0%	NMT 2.0%

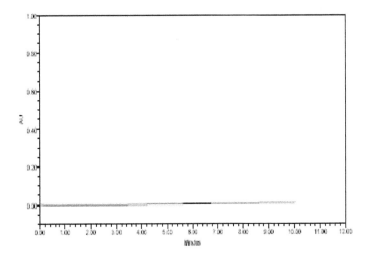

Figure 39: Chromatogram of Blank

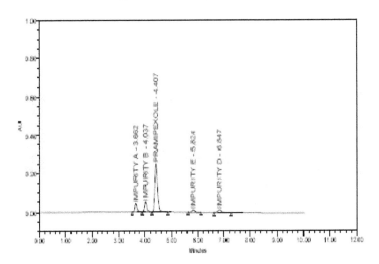

Figure 40 : Resolution chromatogram of Pramipexole and its related Impurities.

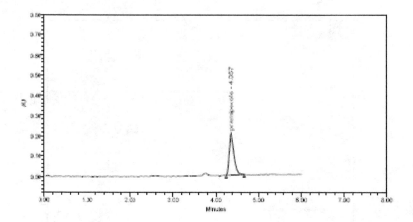

Figure 41: standard 50% solution chromatogram

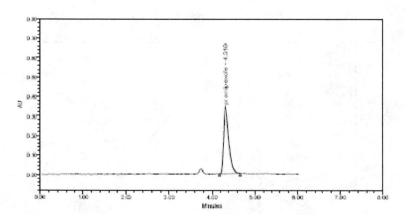

Figure 42: standard 150% solution chromatogram

2.6 Accuracy and Recovery:

Working level concentration of standard solution that is 50,100 & 150 % of assay concentration prepared and injected. System suitability solution, Standard solution and Placebo preparation is as per specificity chapter.

2.6.1 Preparation of Accuracy solution:
Accuracy Stock solution preparation:
Weighed 200 mg of Pramipexole WS into 100 ml of volumetric flask and added 75 ml diluent. Shook well to dissolve and dilute to volume with diluent.

Preparation of 50% solution:
Weighed 100 mg Placebo and added 5 ml of Accuracy std stock solution into 100 ml volumetric flask. Added 75ml of diluent Sonicated for 15 minutes and make up to 100ml with diluent. Filter through 0.45 µm filter paper using syringe. Prepare and inject 50% solutions 3 times separately. The solution is injected and recorded the chromatograms and peak areas.

Preparation of 100% solution:
Weighed 100 mg Placebo and added 10 ml of Accuracy std stock solution into 100 ml volumetric flask. Added 75ml of diluent Sonicated for 15 minutes and make up to 100ml with diluent. Filter through 0.45 µm filter paper using syringe. Prepare and inject 100% solutions 3 times separately. The solution is injected and recorded the chromatograms and peak areas.

Preparation of 150% solution:
Weighed 100 mg Placebo and added 15 ml of Accuracy std stock solution into 100 ml volumetric flask. Added 75ml of diluent Sonicated for 15 minutes and make up to 100ml with diluent. Filter through 0.45 µm filter paper using syringe. Prepare and inject 150% solutions 3 times separately. The solution is injected and recorded the chromatograms and peak areas.

Sys. Suit. was analyzed from resolution solution injection and resolution between peaks of Pramipexole Imp A, Pramipexole Imp B, Pramipexole, Pramipexole Imp E and Pramipexole Imp D. The TF and TP were checked in standard solution. The standard solution for Pramipexole was prepared twice and injected. The parameters such as similarity factor, and RSD of RT, Area of Pramipexole was determined. Results for resolution system suitability are presented in **Table 8 and Table 9**.

Table 8: Result of System suitability-Accuracy (Assay):

Analyte	Resolution
Pramipexole Imp A	N.A.
Pramipexole Imp B	2.2
Pramipexole	2.5
Pramipexole Imp E	7.0
Pramipexole Imp D	4.4

Table 9: Result of System suitability-Accuracy (Assay):

Analyte	% RSD		T.P.	T.F.	Sim. Fac
	RT	Area			
Pramipexole	0.4	0.3	15467	1.3	0.99

The acceptance criterion for similarity factor was ≥0.98 ≤1.02, tailing factor for the peaks due to Pramipexole was NMT 2.0, theoretical plates was not less than 2000 and % RSD of RT, Area of Pramipexole Standard solution of first replicate solution was not more than 1.0% and 2.0% respectively. The acceptance criteria were met.

2.6.2 Procedure:

The working Standard Solution is injected separately 5 times and sample solutions (50% solution, 100% solution & 150% solution) is also injected separately 3 times (about 5µl) respectively.

The % Recovery of Pramipexole is determined by the following Expression.

$$\text{Amount Found} = \frac{\text{Sample area}}{\text{Std Area}} \times \frac{\text{Std wt}}{\text{Std dilution}} \times 1000 \times \frac{\text{Purity}}{100}$$

$$\text{Amount Added} = \frac{\text{Std wt}}{\text{Std dilution}} \times \frac{\text{Purity}}{100} \times 1000$$

The % Recovery is determined by the following Expression.

$$\% \text{ Recovery} = \frac{\text{Amount Found}}{\text{Amount Added}} \times 100$$

The accuracy study of the analytes was determined for the following levels, 50%, 100% and 150% of the specified limit. Analyte showed the recovery between 98% to 102 % concluding the accuracy of the method. The results for both the analytes are presented in **Table 10** which have met the acceptance criteria.

Table 10: Accuracy and recovery of Pramipexole:

Added (µg)	Recovered (µg)	%Recovery (Limit NLT 98.0% and NMT 102.0%)	Mean Recovery	% RSD
100.0	99.5	99.5 %	99.50 %	0.30 %
	99.8	99.8 %		
	99.2	99.2 %		
200.0	202.0	101.0 %	100.85 %	026 %
	201.5	100.75%		
	201.6	100.8 %		
300.0	299.2	99.73 %	99.46 %	1.07 %
	298.7	99.57 %		
	297.2	99.07 %		

At each respective level, the mean recovery for each analyte, from the tablet solution, is within the limit of 98 % - 102 %. Thus, illustrating the HPLC analytical method for the determination of Assay of Pramipexole is accurate.

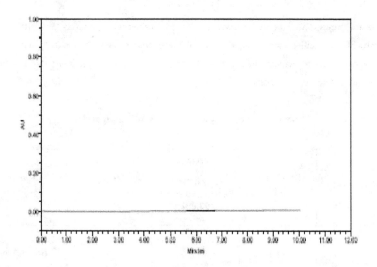

Figure 43: Chromatogram of diluent

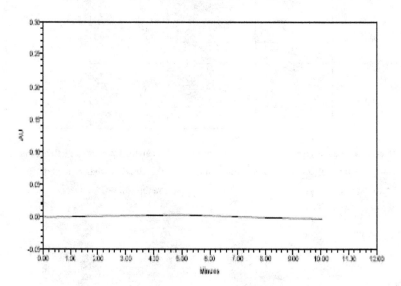

Figure 44: Chromatogram of placebo

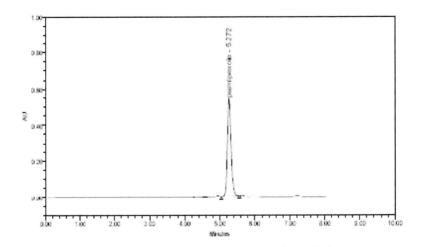

Figure 45: standard solution chromatogram

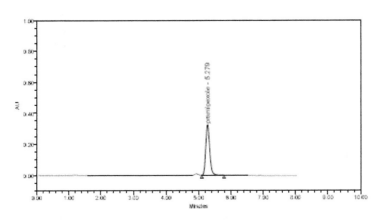

Figure 46: Pramipexole Accuracy 50% Solution chromatogram

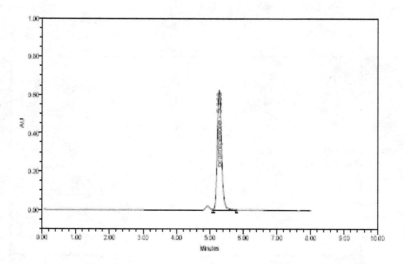

Figure 47: Pramipexole Accuracy 100% Solution chromatogram

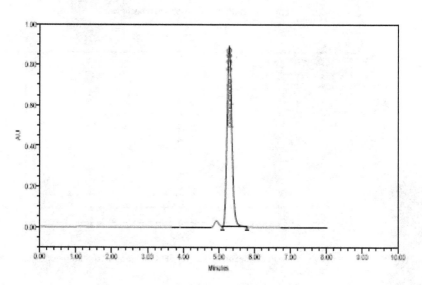

Figure 48: Pramipexole Accuracy 150% Solution chromatogram

2.7 Precision:

System suitability solution, Standard solution, sample solution and Placebo solution preparation is as per specificity chapter.

2.7.1 Procedure:

The working standard solution is injected separately 5 times and six assay sample preparations are injected separately 1 times respectively (5 µl). The solution is injected and recorded the chromatogram and peak areas.

Sys. Suit. was analyzed from resolution solution injection and resolution between peaks of Pramipexole Imp A, Pramipexole Imp B, Pramipexole, Pramipexole Imp E and Pramipexole Imp D. The TF and TP were checked in standard solution. The standard solution for Pramipexole was prepared twice and injected. The parameters such as similarity factor, and RSD of RT , Area of Pramipexole was determined. Results for resolution system suitability are presented in **Table 11 and Table 12**.

Table 11: Result of System suitability-Precision (Assay):

Analyte	Resolution
Pramipexole Imp A	N.A.
Pramipexole Imp B	2.0
Pramipexole	2.4
Pramipexole Imp E	6.9
Pramipexole Imp D	4.5

Table 12: Result of System suitability-Precision (Assay):

Analyte	% RSD		T.P.	T.F.	Sim. Fac
	RT	Area			
Pramipexole	0.5	0.4	14256	1.2	0.99

The acceptance criterion for similarity factor was $\geq 0.98 \leq 1.02$, tailing factor for the peaks due to Pramipexole was NMT 2.0, theoretical plates was not less than 2000 and % RSD of RT ,

Area of Pramipexole Standard solution of first replicate solution was not more than 1.0% and 2.0% respectively. The acceptance criteria were met.

Samples of Pramipexole tablets showed assay value within the limit, hence the precision of the method was done and % RSD of by analyzing sample of pramipexole tablets six times separately. The RSD for recovered Pramipexole was within the limit of 2.00% confirming the precision of the method. **Table 13** represents the method precision results.

Table 13: Method Precision Results of Pramipexole:

Sr. No.	% Recovery of Pramipexole
Sample-1	99.3
Sample-2	99.8
Sample-3	99.0
Sample-4	100.6
Sample-5	97.8
Sample-6	99.6
Mean	99.4
SD	0.93
%RSD	**0.94**

The relative standard deviation of six sample preparation of Pramipexole from the tablet solution was NMT 2.00%. Thus, illustrating the HPLC analytical method is precise.

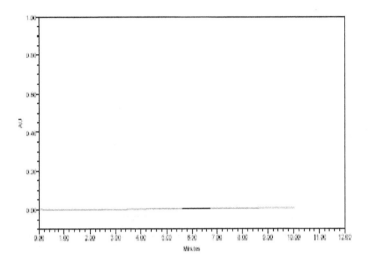

Figure 49: Chromatogram of Blank

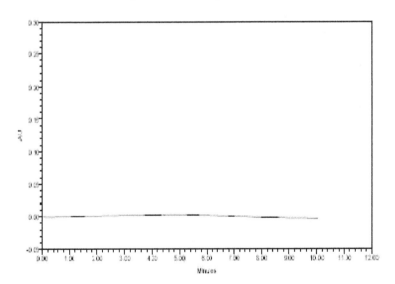

Figure 50: Chromatogram of placebo solution

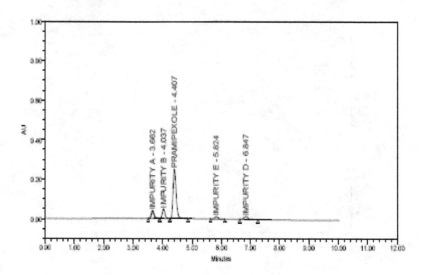

Figure 51 : Resolution chromatogram of Pramipexole and its related Impurities.

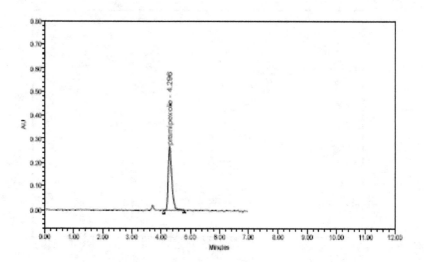

Figure 52: standard solution chromatogram

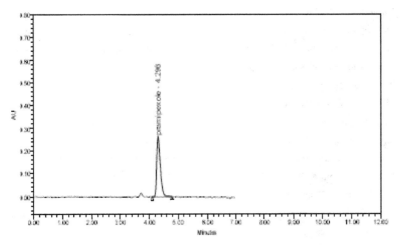

Figure 53: sample solution chromatogram

2.8 Solution Stability:

Stability of solutions in Room temperature from 0th hr, 2nd hr, 4th hr, 6th hr, 8th hr, 12th hr, 16th hr, 20th hr and then up to 24th hrs. System suitability solution, Standard solution, sample solution and Placebo solution preparation is as per specificity chapter.

2.8.1 Procedure:

MP and fresh standard was prepared for system suitability. The working standard solution (freshly prepared) and the assay solution (freshly prepared) is injected separately 5 and 1 replicates respectively and this solution was kept in room temperature from 0hrs to 24th hr.

Sys. Suit. was analyzed from resolution solution injection and resolution between peaks of Pramipexole Imp A, Pramipexole Imp B, Pramipexole, Pramipexole Imp E and Pramipexole Imp D. The TF and TP were checked in standard solution. The standard solution for Pramipexole was prepared twice and injected. The parameters such as similarity factor, and RSD of RT, Area of Pramipexole was determined. Results for resolution system suitability are presented in **Table 14 and Table 15**.

Table 14: Result of System suitability-Solution stability (Assay):

Analyte	Resolution
Pramipexole Imp A	N.A.
Pramipexole Imp B	2.2
Pramipexole	2.3
Pramipexole Imp E	6.6
Pramipexole Imp D	4.7

Table 15: Result of System suitability-Solution stability (Assay):

Analyte	% RSD		T.P.	T.F.	Sim. Fac
	RT	Area			
Pramipexole	0.3	0.7	12890	1.0	0.98

The acceptance criterion for solution stability is same as specificity chapter. The acceptance criteria were met.

The stability of sample solutions were determined at time periods representative for storage. The stability of sample solution was confirmed on the sample by comparing the values of 0th

sample at different time interval to its initial value. The % relative difference should not be more than 2.00. The data for sample solution stability in **Table 16.**

Table 16: Sample solution stability of Pramipexole:

Condition	% Assay of Pramipexole
Sample-0th HR	98.3
Sample-2nd HR	97.6
Sample-4th HR	98.8
Sample-8th HR	97.8
Sample-12th HR	97.6
Sample-16th HR	98.4
Sample-20th HR	98.9
Sample-24th HR	98.1
Mean	98.2
SD	0.51
%RSD	**0.51**

Figure 54: Standard solution stability Curve

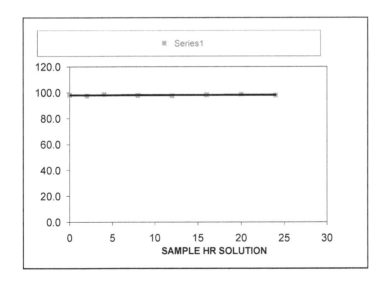

The standard solution was found to be stable till 24 hrs since the % RSD of the peak area response of Pramipexole is not more than 2.0 %. The absolute difference of % assay in sample solution was not more than 2.0 %. so the sample solution is also stable for 24 hrs.

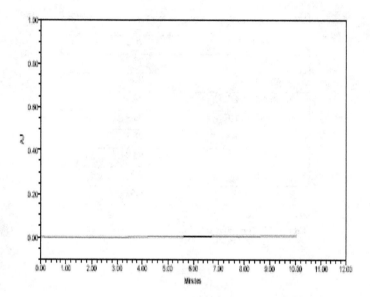

Figure 55: Chromatogram of Blank

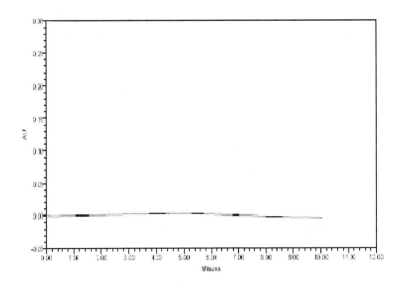

Figure 56: Chromatogram of placebo solution

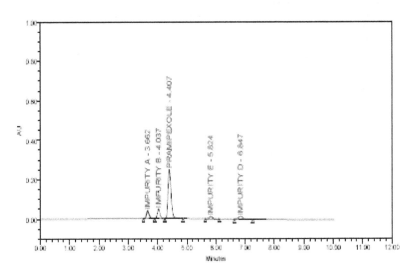

Figure 57: Resolution chromatogram of Pramipexole and its related Impurities.

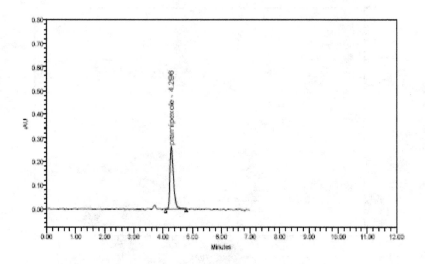

Figure 58: standard solution chromatogram of solution stability

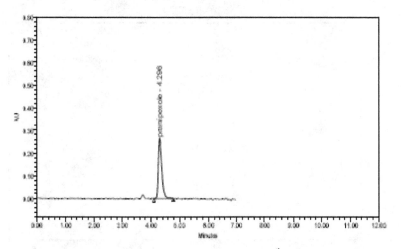

Figure 59: sample solution chromatogram of 0th HR

2.9 Intermediate Precision or Ruggedness:

System suitability solution, Standard solution, sample solution and Placebo solution preparation is as per specificity chapter.

2.9.1 Procedure:

The working standard solution is injected separately 5 times and six assay sample preparations are injected separately 1 times respectively (5 µl). The solution is injected and recorded the chromatogram and peak areas.

Sys. Suit. was analyzed from resolution solution injection and resolution between peaks of Pramipexole Imp A, Pramipexole Imp B, Pramipexole, Pramipexole Imp E and Pramipexole Imp D. The TF and TP were checked in standard solution. The standard solution for Pramipexole was prepared twice and injected. The parameters such as similarity factor, and RSD of RT , Area of Pramipexole was determined. Results for resolution system suitability are presented in **Table 17 and Table 18**.

Table 17: Result of System suitability-Intermediate Precision (Assay):

Analyte	Resolution
Pramipexole Imp A	N.A.
Pramipexole Imp B	2.1
Pramipexole	2.4
Pramipexole Imp E	6.5
Pramipexole Imp D	4.6

Table 18: Result of System suitability-Intermediate Preision (Assay):

Analyte	% RSD		T.P.	T.F.	Sim. Fac
	RT	Area			
Pramipexole	0.2	0.9	13024	0.9	0.99

The acceptance criterion for similarity factor was ≥0.98 ≤1.02, tailing factor for the peaks due to Pramipexole was NMT 2.0, theoretical plates was not less than 2000 and % RSD of RT , Area of Pramipexole Standard solution of first replicate solution was not more than 1.0% and 2.0% respectively. The acceptance criteria were met.

The % RSD was evaluated and NMT 2.0% .The results were presented in **Table 19**.

Table 19: Intermediate Precision Study

Instrument	Waters – Empower
	% Pramipexole
Sr. No.	(A)
1.	99.3
2.	99.8
3.	99.0
4.	100.6
5	97.8
6	99.6
Mean	99.8
SD	1.10
%RSD	1.10

The overall RSD was evaluated and NMT 2.0% .The results were presented in **Table 20**.

Table 20: Cumulative Intermediate Precision Study

Instrument	Waters – Empower	
	% Pramipexole	
Sr. No.	(A)	(B)
1.	99.3	100.6
2.	99.8	101.2
3.	99.0	100.1
4.	100.6	99.7
5	97.8	99.1
6	99.6	98.1
Mean	99.6	
SD	1.00	
%RSD	1.01	

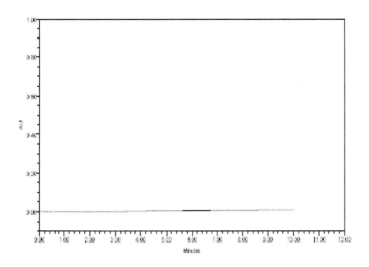

Figure 60: Chromatogram of Blank

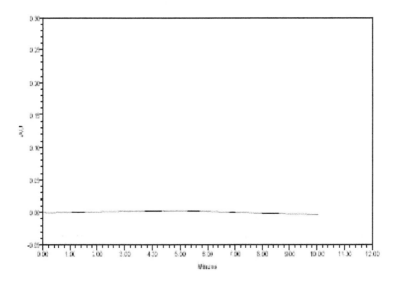

Figure 61: Chromatogram of placebo solution

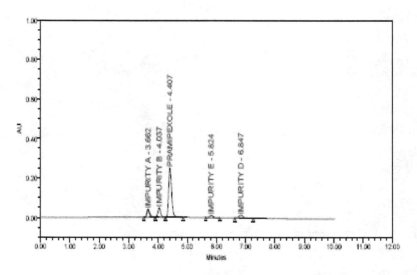

Figure 62 : Resolution chromatogram of Pramipexole and its related Impurities.

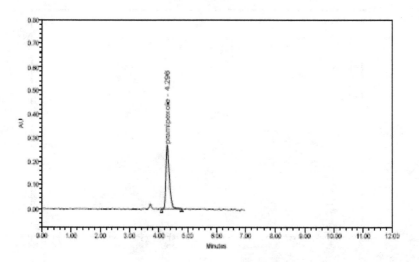

Figure 63: standard solution chromatogram of Intermediate precision

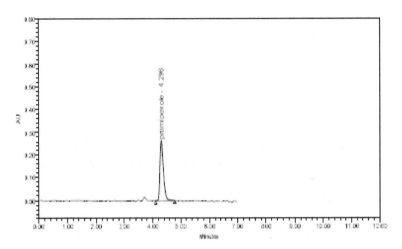

Figure 64: sample solution chromatogram of Intermediate precision

2.10 Robustness:

In all the deliberate varied chromatographic conditions carried out (MP flow rate and column temperature in the variants),

2.10.1 Robustness-I: (Flow rate changed by –0.1ml)

System suitability solution, Standard solution, sample solution and Placebo solution preparation is as per specificity chapter.

2.10. 2 Procedure:

The working standard solution is injected separately 5 times and three assay sample preparations are injected separately 1 times (5 µl). The solution is injected and recorded the chromatogram and peak areas.

Sys. Suit. was analyzed from resolution solution injection and resolution between peaks of Pramipexole Imp A, Pramipexole Imp B, Pramipexole, Pramipexole Imp E and Pramipexole Imp D. The TF and TP were checked in standard solution. The standard solution for Pramipexole was prepared twice and injected. The parameters such as similarity factor, and RSD of RT , Area of Pramipexole was determined. Results for resolution system suitability are presented in **Table 21 and Table 22**.

Table 21: Result of System suitability-Robustness - I (Assay):

Analyte	Resolution
Pramipexole Imp A	N.A.
Pramipexole Imp B	2.3
Pramipexole	2.2
Pramipexole Imp E	6.7
Pramipexole Imp D	4.6

Table 22: Result of System suitability-Robustness - I (Assay):

Analyte	% RSD		T.P.	T.F.	Sim. Fac
	RT	Area			
Pramipexole	0.5	1.4	13856	1.2	1.00

The acceptance criterion for similarity factor was ≥0.98 ≤1.02, tailing factor for the peaks due to Pramipexole was NMT 2.0, theoretical plates was not less than 2000 and % RSD of RT ,

Area of Pramipexole Standard solution of first replicate solution was not more than 1.0% and 2.0% respectively. The acceptance criteria were met.

The resolution does not show significant changes in robustness. This shows minor change of the parameters does not affect the chromatographic separation.

It was determined on three separate sample solutions prepared from same batch. The % RSD was evaluated and NMT 2.0% .The results were presented in **Table 23**.

Table 23: Robustness-I (Flow rate changed by –0.1ml) Assay of Pramipexole

Sample No.	Assay of Pramipexole
1	96.0
2	97.1
3	96.5
Mean	96.5
SD	0.55
%RSD	0.57

The cumulative % RSD was evaluated with the method precision samples and NMT 2.0%. The results were presented in **Table 24**.

Table 24: Robustness-I (Flow rate changed by –0.1ml) cumulative %RSD of Pramipexole

Sample No.	Assay of Pramipexole
1	99.3
2	99.8
3	99.0
4	100.6
5	97.8
6	99.6
7	96.0
8	97.1
9	96.5
Mean	98.4
SD	1.61
%RSD	1.64

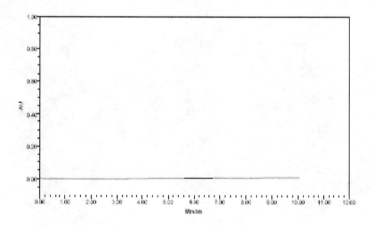

Figure 65: Chromatogram of Blank

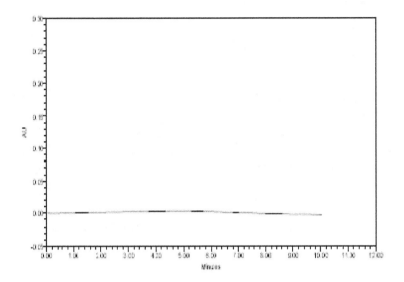

Figure 66: Chromatogram of placebo solution

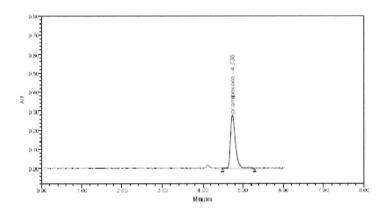

Figure 67: standard solution chromatogram of Rob I

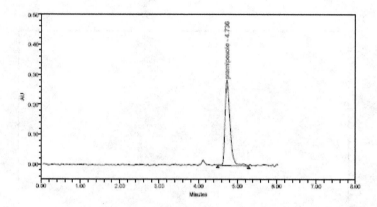

Figure 68: sample solution chromatogram of Rob I

2.10.3 Robustness -II:(Flow rate changed by +0.1ml)
(Resolution standard solution, Standard solution, Sample solution (three preparations and placebo solution prepared as robustness-I chapter)

2.10.4 Procedure:
The working standard solution is injected separately 5 times and three assay sample preparations are injected separately 1 times (5 µl). The solution is injected and recorded the chromatogram and peak areas.

Sys. Suit. was analyzed from resolution solution injection and resolution between peaks of Pramipexole Imp A, Pramipexole Imp B, Pramipexole, Pramipexole Imp E and Pramipexole Imp D. The TF and TP were checked in standard solution. The standard solution for Pramipexole was prepared twice and injected. The parameters such as similarity factor, and RSD of RT, Area of Pramipexole was determined. Results for resolution system suitability are presented in **Table 25 and Table 26**.

Table 25: Result of System suitability- Robustness-II (Assay):

Analyte	Resolution
Pramipexole Imp A	N.A.
Pramipexole Imp B	2.2
Pramipexole	2.3
Pramipexole Imp E	6.8
Pramipexole Imp D	4.7

Table 26: Result of System suitability- Robustness-II (Assay):

Analyte	% RSD		T.P.	T.F.	Sim. Fac
	RT	Area			
Pramipexole	0.2	0.4	13408	1.1	1.01

The acceptance criterion for similarity factor was ≥0.98 ≤1.02, tailing factor for the peaks due to Pramipexole was NMT 2.0, theoretical plates was not less than 2000 and % RSD of RT, Area of Pramipexole Standard solution of first replicate solution was not more than 1.0% and 2.0% respectively. The acceptance criteria were met.

The resolution does not show significant changes in robustness. This shows minor change of the parameters does not affect the chromatographic separation.

It was determined on three separate sample solutions prepared from same batch. The % RSD was evaluated and NMT 2.0% .The results were presented in **Table 27**.

Table 27: Robustness-II (Flow rate changed by +0.1ml) Assay of Pramipexole

Sample No.	Assay of Pramipexole
1	101.1
2	100.5
3	99.8
Mean	100.5
SD	0.65
%RSD	0.65

The cumulative % RSD was evaluated with the method precision samples and were within the acceptance criterion of NMT 2.0% .The results were presented in **Table 28**.

Table 28: Robustness-II (Flow rate changed by +0.1ml) cumulative %RSD of Pramipexole

Sample No.	Assay of Pramipexole
1	99.3
2	99.8
3	99.0
4	100.6
5	97.8
6	99.6
7	101.1
8	100.5
9	99.8
Mean	99.7
SD	0.98
%RSD	0.98

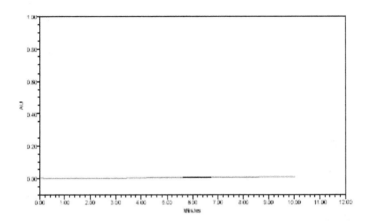

Figure 69: Chromatogram of Blank

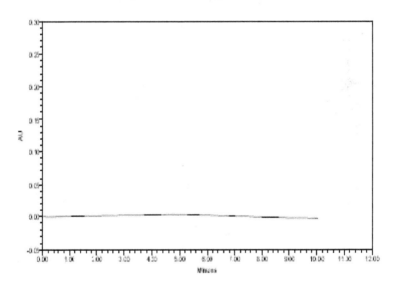

Figure 70: Chromatogram of placebo solution

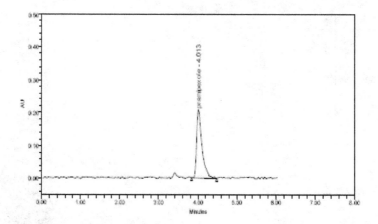

Figure 71: standard solution chromatogram of Rob II

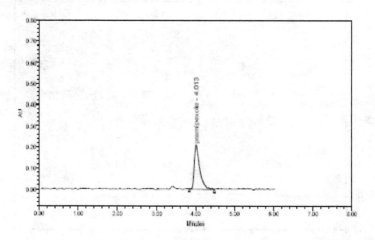

Figure 72: sample solution chromatogram of Rob II

2.10.5 Robustness -III: (Column oven Temp. changed by +2)
(Resolution standard solution, Standard solution, Sample solution (three preparations and placebo solution prepared as robustness-I chapter)

2.10.6 Procedure:

The working standard solution is injected separately 5 times and three assay sample preparations are injected separately 1 times (5 µl). The solution is injected and recorded the chromatogram and peak areas.

Sys. Suit. was analyzed from resolution solution injection and resolution between peaks of Pramipexole Imp A, Pramipexole Imp B, Pramipexole, Pramipexole Imp E and Pramipexole Imp D. The TF and TP were checked in standard solution. The standard solution for Pramipexole was prepared twice and injected. The parameters such as similarity factor, and RSD of RT, Area of Pramipexole was determined. Results for resolution system suitability are presented in **Table 29 and Table 30**.

Table 29: Result of System suitability- Robustness-III (Assay):

Analyte	Resolution
Pramipexole Imp A	N.A.
Pramipexole Imp B	2.1
Pramipexole	2.4
Pramipexole Imp E	6.5
Pramipexole Imp D	4.8

Table 30: Result of System suitability- Robustness-III (Assay):

Analyte	% RSD		T.P.	T.F.	Sim. Fac
	RT	Area			
Pramipexole	0.4	1.2	12609	1.3	1.00

The acceptance criterion for similarity factor was ≥0.98 ≤1.02, tailing factor for the peaks due to Pramipexole was NMT 2.0, theoretical plates was not less than 2000 and % RSD of RT, Area of Pramipexole Standard solution of first replicate solution was not more than 1.0% and 2.0% respectively. The acceptance criteria were met.

The resolution does not show significant changes in robustness. This shows minor change of the parameters does not affect the chromatographic separation.

It was determined on three separate sample solutions prepared from same batch. The % RSD was evaluated and were within the acceptance criterion of NMT 2.0% .The results were presented in **Table 31**.

Table 31: Robustness-III (Column oven Temp. changed by +2) Assay of Pramipexole

Sample No.	Assay of Pramipexole
1	96.1
2	96.9
3	97.9
Mean	97.0
SD	0.90
%RSD	0.93

The cumulative % RSD was evaluated with the method precision samples and were within the acceptance criterion of NMT 2.0% .The results were presented in **Table 32**.

Table 32: Robustness-III (Column oven Temp. changed by +2) cumulative %RSD of Pramipexole

Sample No.	Assay of Pramipexole
1	99.3
2	99.8
3	99.0
4	100.6
5	97.8
6	99.6
7	96.1
8	96.9
9	97.9
Mean	98.6
SD	1.47
%RSD	1.49

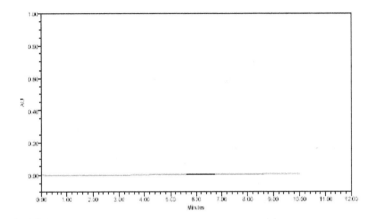

Figure 73: Chromatogram of Blank

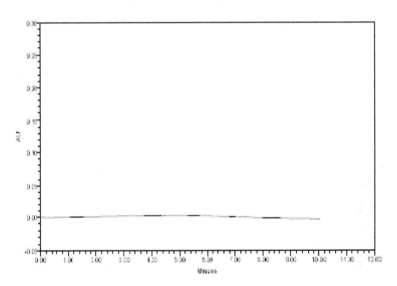

Figure 74: Chromatogram of placebo solution

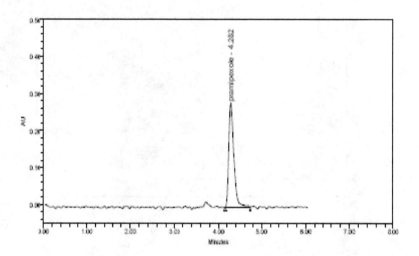

Figure 75: standard solution chromatogram of Rob III

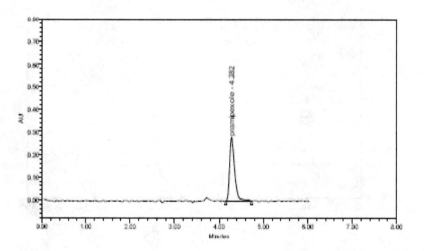

Figure 76: sample solution chromatogram of Rob III

2.10.7 Robustness -IV :(Column oven Temp. changed by -2)
(Resolution standard solution, Standard solution, Sample solution (three preparations and placebo solution prepared as robustness-I chapter)
2.10.8 Procedure:
The working standard solution is injected separately 5 times and three assay sample preparations are injected separately 1 times (5 µl). The solution is injected and recorded the chromatogram and peak areas.

Sys. Suit. was analyzed from resolution solution injection and resolution between peaks of Pramipexole Imp A, Pramipexole Imp B, Pramipexole, Pramipexole Imp E and Pramipexole Imp D. The TF and TP were checked in standard solution. The standard solution for Pramipexole was prepared twice and injected. The parameters such as similarity factor, and RSD of RT , Area of Pramipexole was determined. Results for resolution system suitability are presented in **Table 33 and Table 34**.

Table 33: Result of System suitability- Robustness-IV (Assay):

Analyte	Resolution
Pramipexole Imp A	N.A.
Pramipexole Imp B	2.0
Pramipexole	2.3
Pramipexole Imp E	6.6
Pramipexole Imp D	4.6

Table 34: Result of System suitability- Robustness-IV (Assay):

Analyte	% RSD		T.P.	T.F.	Sim. Fac
	RT	Area			
Pramipexole	0.2	0.3	12589	1.2	0.99

The acceptance criterion for similarity factor was ≥0.98 ≤1.02, tailing factor for the peaks due to Pramipexole was NMT 2.0, theoretical plates was not less than 2000 and % RSD of RT , Area of Pramipexole Standard solution of first replicate solution was not more than 1.0% and 2.0% respectively. The acceptance criteria were met.

The resolution does not show significant changes in robustness. This shows minor change of the parameters does not affect the chromatographic separation.

It was determined on three separate sample solutions prepared from same batch. The % RSD was evaluated and NMT 2.0% .The results were presented in **Table 35**.

Table 35: Robustness-IV (Column oven Temp. changed by -2) cumulative %RSD of Pramipexole

Sample No.	Assay of Pramipexole
1	102.1
2	101.8
3	100.9
Mean	101.6
SD	0.62
%RSD	0.61

The cumulative % RSD was evaluated with the method precision samples and NMT 2.0% .The results were presented in **Table 36**.

Table 36: Robustness-IV (Column oven Temp. changed by -2) cumulative %RSD of Pramipexole

Sample No.	Assay of Pramipexole
1	99.3
2	99.8
3	99.0
4	100.6
5	97.8
6	99.6
7	102.1
8	101.8
9	100.9
Mean	100.1
SD	1.38
%RSD	1.38

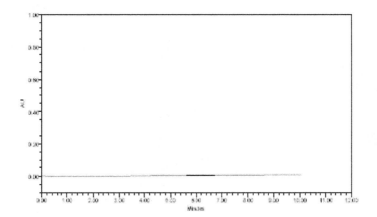

Figure 77: Chromatogram of Blank

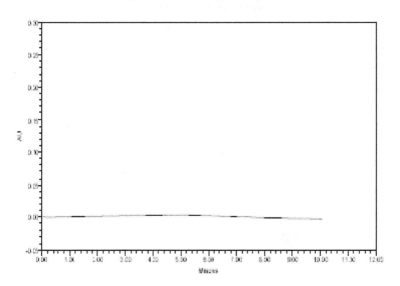

Figure 78: Chromatogram of placebo solution

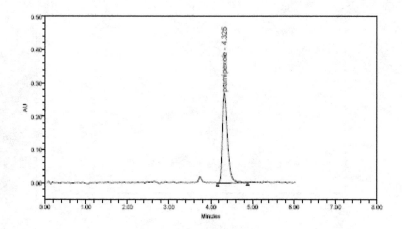

Figure 79: standard solution chromatogram of Rob IV

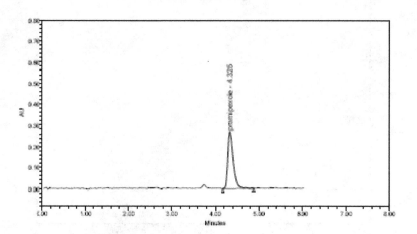

Figure 80: sample solution chromatogram of Rob IV

Chapter 3
Related substances
Method Validation

3.1 Specificity and System Suitability:

3.1.1 Specificity:

The methods validation parameter specificity is determined by comparing chromatograms of diluent, actives, the related impurities and the placebo (containing all the ingredients of the formulation except the analytes) of the tablets as per the procedure applied to sample solution.

3.1.2 Preparation of Resolution solution for Related substances:
Preparation of Imp stock solution:
Weighed 1.5 mg of each Pramipexole Imp standard (Pramipexole A, Pramipexole B, Pramipexole D and Pramipexole E) into 10 ml of volumetric flask added 7 ml of diluent sonicated to dissolved and diluted to volume with diluent. (concentration of each Imp is 150 µg/ml)

Preparation of Resolution solution :
Weighed 20 mg of Pramipexole standard and into this added 1 ml of Imp stock solution into 100 ml of volumetric flask added 70 ml of diluent sonicated to dissolved and diluted to volume with diluent. (Standard concentration is 200 µg/ml and each Imp concentration is 1.5 µg/ml)

3.1.3 Preparation of Standard solution:
Standard Preparation: Weighed 20 mg of Pramipexole standard into 100 ml of volumetric flask added 70 ml of diluent sonicated to dissolved and diluted to volume with diluent.
(Standard concentration is 200 µg/ml) .
Further diluted 1 ml of std stock solution and 1 ml of Imp stock solution in to 100 ml volumetric flask and diluted to volume with diluents. (Standard concentration is 2 µg/ml and each Imp concentration is 1.5 µg/ml)
The standard solution prepared twice and injected for checking the similarity factor. Then the standard solution A was injected 6 times separately. The solution is injected and recorded the chromatograms and peak areas.

3.1.4 Preparation of Sample solution:
Sample Preparation: Weighed sample equivalent to 20 mg of Pramipexole into 20 ml of volumetric flask added 15 ml of diluent sonicated to dissolved and diluted to volume with diluent. Filter through 0.45 µm filter paper using syringe. The solution is injected and recorded the chromatograms and peak areas.
(Sample concentration is 1000 µg/ml).

3.1.5 Placebo Preparation:
An accurately weighed portion of the placebo, equivalent to 20 mg of Pramipexole was transferred into a 20 ml volumetric flask containing 10 ml of diluent and disperses with the aid of ultrasound for 10 minutes with intermittent swirling. The flask was further shaken with the means of mechanical shaker for 15 minutes and allowed to reach the ambient room temperature. The volume was made up to 20 ml with diluent and mixed. Filtered the solution through GF/C.

System suitability was performed by injecting resolution solution and determining resolution between closely eluting peak of Pramipexole, Pramipexole A, Pramipexole B, Pramipexole D and Pramipexole E. The TF and TP were checked in Standard solution. The standard solution for Pramipexole was prepared twice and injected. The parameters such as similarity factor, and RSD of RT , Area of Pramipexole was determined. Results for resolution system suitability are presented in **Table 37**.

Table 37: Result of System suitability-Specificity(Related substances):

Analyte	Resolution
Pramipexole Imp A	N.A.
Pramipexole Imp B	2.2
Pramipexole	2.5
Pramipexole Imp E	6.8
Pramipexole Imp D	4.6

Table 38: Result of System suitability-Specificity(Related substances):

Analyte	% RSD		T.P.	T.F.	Sim. Fac
	RT	Area			
Pramipexole Imp A	0.8	1.5	8628	1.2	0.98
Pramipexole Imp B	0.6	1.8	10108	1.0	0.96
Pramipexole	0.7	1.1	8396	1.0	0.98
Pramipexole Imp E	0.5	1.7	12442	1.2	0.97
Pramipexole Imp D	0.5	1.1	12500	1.1	0.96

The acceptance criterion for similarity factor was ≥0.98 ≤1.02, tailing factor for the peak due to Pramipexole, Pramipexole A, Pramipexole B, Pramipexole D and Pramipexole E. was NMT 2.0, theoretical plates was not less than 2000 and % RSD of RT , Area of Standard solution of first replicate solution was not more than 1.0% and 5.0% respectively. The acceptance criteria were met.

Calculations:

Calculate the % known and unknown Imp/impurities in the Chromatogram obtained with Sample solution as follows,

a) Known Imp = $\dfrac{A_T}{A_S} \times \dfrac{D_S}{D_T} \times \dfrac{P}{100} \times \dfrac{100}{L} \times N \times \dfrac{1}{RRF}$

b) Unknown Imp = $\dfrac{A_{T1}}{A_S} \times \dfrac{D_S}{D_T} \times \dfrac{P}{100} \times \dfrac{100}{L} \times N$

c) Total Impurities = Sum of all known and unknown impurities

Where,

A_T = Peak area of known Imp in the chromatogram of Sample solution.

A_{T1} = Peak area of individual unknown Imp in the chromatogram of Sample solution.
A_S = Average Peak area of Pramipexolein the chromatogram of diluted standard solution
D_S = Dilution factor for the standard solution (weight ÷ dilution).
D_T = Dilution factor for the sample solution (weight ÷ dilution).
P = Percent potency of working standard used (as is basis)
L = Label claim of Pramipexolein mg.
N = Average weight of the tablet in mg.
RRF = Relative response factor

Note:
(1) If the known / unknown Imp is not detected; report as not detected.
(2) If known / unknown Imp is equal to LOQ value; report the value and add in total impurities calculation.
(3) If known / unknown Imp is less than LOQ value; report as BLOQ and do not add in total impurities calculation.

Diluent (**Figure 81**), Placebo Preparation (**Figure 82**), Resolution solution and standard Preparation (**Figure 83**), were injected into the chromatograph. No interference was detected at the RT of Pramipexole and their respective impurities hence proving the specificity of the method. **Table 39** indicate the relative retention times (RRT) of all the impurities.

Table 39: Identification Study of Related substances

Sr. No	Components	Retention Time (mins)	Relative retention time (RRT)	Placebo peak area response
1	Pramipexole Imp A	3.100	0.69	Not detected
2	Pramipexole Imp B	4.102	0.91	Not detected
3	Pramipexole	4.516	1.00	Not detected
4	Pramipexole Imp E	5.654	1.25	Not detected
5	Pramipexole Imp D	6.725	1.49	Not detected

There were no peaks due to the MP, Diluent and Placebo at the retention time of Pramipexole and their impurities. This shows that method is specific by HPLC determination.

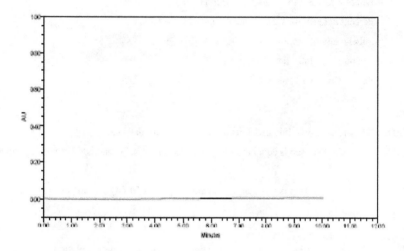

Figure 81: Chromatogram of Blank

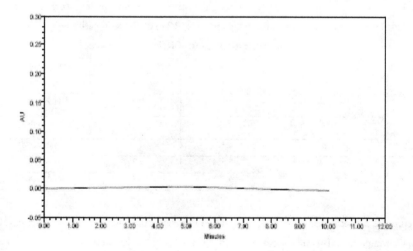

Figure 82: Chromatogram of placebo

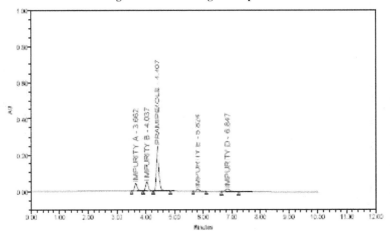

Figure 83 : Resolution chromatogram of Pramipexole and its related Impurities.

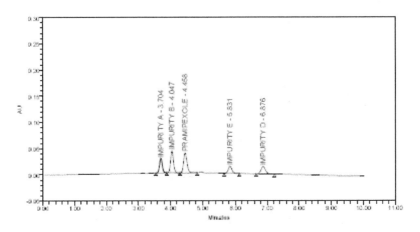

Figure 84: standard solution chromatogram

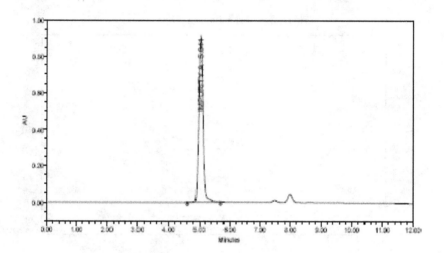

Figure 85: Pramipexole Imp A Solution chromatogram

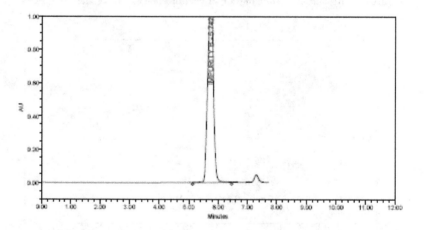

Figure 86: Pramipexole Imp B Solution chromatogram

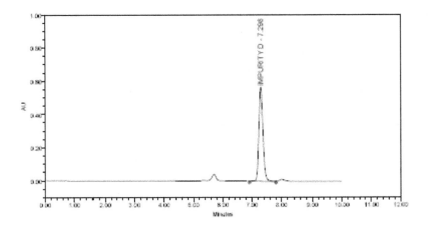

Figure 87: Pramipexole Imp D Solution chromatogram

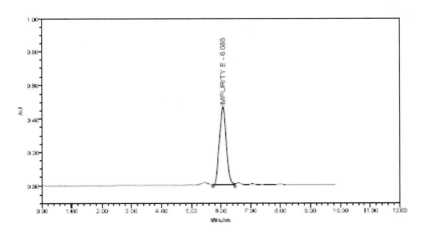

Figure 88: Pramipexole Imp E Solution chromatogram

3.2 Determination of Limit of Detection :

The detection limit of an individual analytical procedure is the lowest amount of analyte or known impurities in a sample which can be detected but not quantified.
Limit of Detection is to be determined from linearity plots of Relative Response Factors.

3.2.1 Formula for prediction of Limit of detection (LOD) :

$$LOD = 3.3 \times \frac{\sigma}{Slope}$$

Where,

σ : Residual standard deviation of the response

Slope: Slope to be taken from the Linearity plot of the component for which the LOD is to be determined.

Table No.40: Summary for Signal to noise ration of peak obtained during LOD Chapter.

Sr. No.	Signal to Noise ratio obtained with LOD solution for				
	Pramipexole	Pramipexole Imp A	Pramipexole (Imp D)	Pramipexole Imp B	Pramipexole Imp E
1	5	3	6	8	6
2	6	5	5	8	7
3	5	4	8	7	9
Avg.	5	4	6	8	7

Figure 89: LOD solution of Pramipexole Imp chromatogram

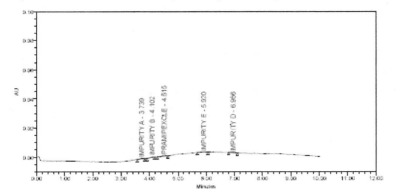

3.3 Determination of Limit of Quantitation:

After determination of LOQ level concentration, inject LOQ solutions for Pramipexoleas well as each known Imp for six replicates and carry out the precision activity for this LOQ solution. Limit of Quantitation is to be determined from linearity plots of Relative Response Factors

3.3.1 Formula for prediction of Limit of Quantitation (LOQ):

$$LOD = 10.0 \times \frac{\sigma}{Slope}$$

Where,

σ : Residual standard deviation of the response

Slope: Slope to be taken from the Linearity plot of the component for which the LOQ is to be determined.

Table No. 41 : Prediction of LOD and LOQ levels w.r.t. working concentration :

Component Name	ppm Limit of detection	ppm Limit of quantitation
Pramipexole	0.054	0.75
Pramipexole Imp A	0.036	0.50
Pramipexole Imp D	0.032	0.50
Pramipexole Imp B	0.058	0.50
Pramipexole Imp E	0.095	0.50

Table No.42: Summary for Signal to noise ration of peak obtained during LOQ Study.

Sr. No.	Signal to Noise ratio obtained with LOQ solution for				
	Pramipexole	Pramipexole Imp A	Pramipexole(Imp D)	Pramipexole Imp B	Pramipexole Imp E
1	12	14	13	14	23
2	13	11	16	17	21
3	12	14	15	11	26
4	14	14	17	16	20
5	11	11	18	17	21
6	11	14	16	16	23
Avg.	12	13	16	15	22

Table No.43: Results summary for RSD of peak areas obtained during LOQ Study.

Sr. No.	Parameter	Results	Acceptance criteria
1	RSD of Area of Pramipexole	0.1	Not more than 5.0 %
2	RSD of Area of Pramipexole Imp A	0.4	Not more than 5.0 %
3	RSD of Area of Pramipexole Imp D	0.6	Not more than 5.0 %
4	RSD of Area of Pramipexole Imp B	0.3	Not more than 5.0 %
5	RSD of Area of Pramipexole Imp E	0.8	Not more than 5.0 %

Figure 90: LOQ solution of Pramipexole Imp chromatogram

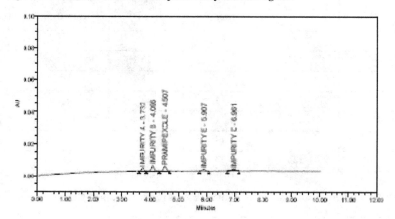

Conclusion : As RSD for area of LOQ level for Pramipexole and each known impurities are well within limit. Hence method is precise at LOQ level.

3.4 Linearity and Range:

The concentration of serial standard solution is LOQ, 50, 75, 100, 125 & 150% of limit concentration (working level 2 ppm for Pramipexole) was prepared and injected.

3.4.1 Linearity solution preparation:

Preparation of Imp stock solution:

Weighed 1.5 mg of each Pramipexole Imp standard (Pramipexole A, Pramipexole B, Pramipexole D and Pramipexole E) into 10 ml of volumetric flask added 7 ml of diluent sonicated to dissolved and diluted to volume with diluent. (concentration of each Imp is 150 µg/ml)

Linearity Stock solution preparation:

Weighed 20 mg of Pramipexole WS into 100 ml of volumetric flask and added 75 ml diluent. Shook well to dissolve and dilute to volume with diluent.

Preparation of LOQ solution:

Pipetted out 0.375 ml of std stock solution of linearity and 0.33 ml of Imp stock solution in to 100 ml volumetric flask and diluted to volume with diluents. (Standard concentration is 0.75 µg/ml and each Imp concentration is 0.5 µg/ml).

Then the solution was injected 6 times separately. The solution is injected and recorded the chromatograms and peak areas.

Preparation of 50% solution:
Pipetted out 0.5 ml of std stock solution of linearity and 0.5 ml of Imp stock solution in to 100 ml volumetric flask and diluted to volume with diluents. (Standard concentration is 1 µg/ml and each Imp concentration is 0.75 µg/ml)
Then the solution was injected 2 times separately. The solution is injected and recorded the chromatograms and peak areas.

Preparation of 75% solution:
Pipetted out 0.75 ml of std stock solution of linearity and 0.75 ml of Imp stock solution in to 100 ml volumetric flask and diluted to volume with diluents. (Standard concentration is 1.5 µg/ml and each Imp concentration is 1.13 µg/ml)
Then the solution was injected 2 times separately. The solution is injected and recorded the chromatograms and peak areas.

Preparation of 100% solution:
Pipetted out 1 ml of std stock solution of linearity and 1 ml of Imp stock solution in to 100 ml volumetric flask and diluted to volume with diluents. (Standard concentration is 2 µg/ml and each Imp concentration is 1.5 µg/ml).
Then the solution was injected 2 times separately. The solution is injected and recorded the chromatograms and peak areas.

Preparation of 125% solution:
Pipetted out 1.25 ml of std stock solution of linearity and 1.25 ml of Imp stock solution in to 100 ml volumetric flask and diluted to volume with diluents. (Standard concentration is 2.5 µg/ml and each Imp concentration is 1.88 µg/ml).
Then the solution was injected 2 times separately. The solution is injected and recorded the chromatograms and peak areas.

Preparation of 150% solution:
Pipetted out 1.5 ml of std stock solution of linearity and 1.5 ml of Imp stock solution in to 100 ml volumetric flask and diluted to volume with diluents. (Standard concentration is 3.0 µg/ml and each Imp concentration is 2.25 µg/ml).

Then the solution was injected 6 times separately. The solution is injected and recorded the chromatograms and peak areas.

3.4.2 Procedure:

First level (low conc. LOQ solution) and Last level (high conc. 150% solution) is injected separately 6 times and other levels (exception of low & high conc.) are injected separately 2 times (5 µl).

The chromatogram is to be recorded and measured the peak response.

Calculation:

% Y-Intercept:

$$\% \text{ Y-Intercept} = \frac{\text{Y-intercept}}{\text{Mean area of working level conc.}} \times 100$$

Response Factor:

$$\text{Response Factor} = \frac{\text{Mean Response of each Linearity}}{\text{Concentration of Respective linearity level}}$$

A series of Solutions of standard Pramipexole was prepared over the range of LOQ level to 150% level of the specified limit for Related substances. The results obtained are presented in **Table 44 and Table 48**.

Fig. 91: Linearity Plot of Pramipexole for Linearity Study.

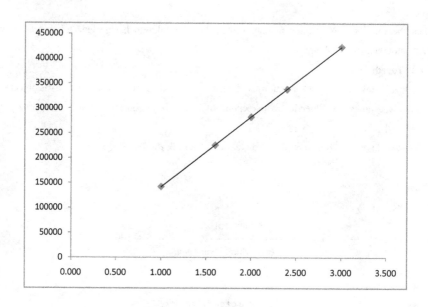

Table No. 44: Results of Linearity of Pramipexole.

Sr. No.	Parameter	Results	Acceptance criteria
1	Correlation coefficient (R)	0.9999	≥ 0.99
2	% Y - Intercept	0.44	$\leq \pm 25\%$
3	Slope of regression	140517	To be reported

Fig. 92: Linearity Plot of Pramipexole Imp A for RRF Study.

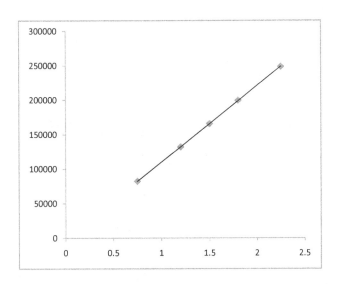

Table No. 45: Results of Linearity of Pramipexole Imp A

Sr. No.	Parameter	Results	Acceptance criteria
1	Correlation coefficient (R)	0.9999	≥ 0.99
2	% Y - Intercept	0.7	$\leq \pm 25\%$
3	Slope of regression	1665	To be report only

Fig. 93: Linearity Plot of Pramipexole Imp D for Linearity Study.

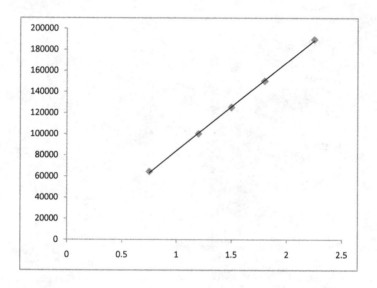

Table No. 46: Results of Linearity of Pramipexole Imp D

Sr. No.	Parameter	Results	Acceptance criteria
1	Correlation coefficient (R)	1.0000	≥ 0.99
2	% Y - Intercept	0.4	$\leq \pm 25.0\%$
3	Slope of regression	83402	To be report only

Fig. 94: Linearity Plot of Pramipexole Imp B for Linearity Study.

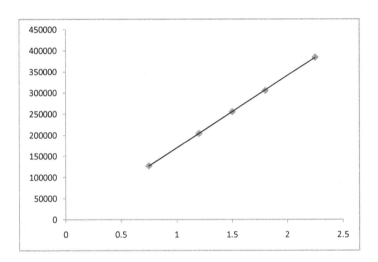

Table No. 47: Results of Linearity of Pramipexole Imp B :

Sr. No.	Parameter	Results	Acceptance criteria
1	Correlation coefficient (R)	0.9999	≥ 0.99
2	% Y - Intercept	-0.1	≤ ± 25.0 %
3	Slope of regression	170937	To be report only

Fig. 95: Linearity Plot of Pramipexole Imp E for Linearity Study.

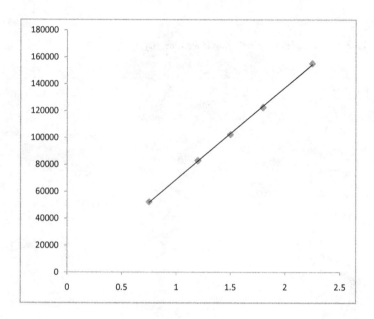

Table No. 48: Results of Linearity of Pramipexole Imp E

Parameter	Results	Acceptance criteria
Correlation coefficient (R)	0.9996	≥ 0.99
% Y - Intercept	0.0	≤ ± 25.0 %
Slope of regression	68287	To be report only

3.5 Accuracy :

Accuracy of an analytical procedure expresses the closeness of agreement between the value which is accepted either as a conventional true value or an accepted reference value and the value found.

For accuracy study of known Imp, specified level of impurities shall be spiked in sample solution while for accuracy study of Pramipexole, specified level of drug substance solutions shall be spiked in the sample.

Accuracy shall be determined on 15 determinations over a minimum of three concentration levels covering the specified range (i.e. LOQ to 150%).

Accuracy study shall be carried out for Pramipexole as well as for known Imp as follows.

3.5.1 Preparation of Imp stock solution:

Weighed 1.5 mg of each Pramipexole Imp standard (Pramipexole A, Pramipexole B, Pramipexole D and Pramipexole E) into 10 ml of volumetric flask added 7 ml of diluent sonicated to dissolved and diluted to volume with diluent. (Concentration of each Imp is 150 µg/ml)

3.5.2 Sample Preparation:

Take 20 tablets, crush finely to powder and weigh powder equivalent to 20 mg of Pramipexole in a 20.0 mL dry volumetric flask. Add about 10 mL of diluent. Further sonicate for 10 minutes with intermittent shaking, , cool and dilute to volume with diluent. Filter through Nylon or PVDF filter 0.2 µm after discarding first 8.0 mL of filtrate.

3.5.3 Preparation of LOQ Spike sample :

Take 20 tablets, crush finely to powder and weigh powder equivalent to 20 mg of Pramipexole in a 20.0 mL dry volumetric flask. Further diluted 0.066 ml of Known Imp standard stock solution of Imp A, Imp B, Imp D and Imp E in to volumetric flask and Add about 10 mL of diluent. Further sonicate for 10 minutes with intermittent shaking, cool and dilute to volume with diluent. Filter through Nylon or PVDF filter 0.2 µm after discarding first 8.0 mL of filtrate.

3.5.4 Preparation of 50 % Spike sample :
Take 20 tablets, crush finely to powder and weigh powder equivalent to 20 mg of Pramipexole in a 20.0 mL dry volumetric flask. Further diluted 0.10 ml of Known Imp standard stock solution of Imp A, Imp B, Imp D and Imp E in to volumetric flask and Add about 10 mL of diluent. Further sonicate for 10 minutes with intermittent shaking, cool and dilute to volume with diluent. Filter through Nylon or PVDF filter 0.2 µm after discarding first 8.0 mL of filtrate.

3.5.5 Preparation of 100 % Spike sample :
Take 20 tablets, crush finely to powder and weigh powder equivalent to 20 mg of Pramipexole in a 20.0 mL dry volumetric flask. Further diluted 0.20 ml of Known Imp standard stock solution of Imp A, Imp B, Imp D and Imp E in to volumetric flask and Add about 10 mL of diluent. Further sonicate for 10 minutes with intermittent shaking, cool and dilute to volume with diluent. Filter through Nylon or PVDF filter 0.2 µm after discarding first 8.0 mL of filtrate.

3.5.6 Preparation of 150 % Spike sample :
Take 20 tablets, crush finely to powder and weigh powder equivalent to 20 mg of Pramipexole in a 20.0 mL dry volumetric flask. Further diluted 0.30 ml of Known Imp standard stock solution of Imp A, Imp B, Imp D and Imp E in to volumetric flask and Add about 10 mL of diluent. Further sonicate for 10 minutes with intermittent shaking, cool and dilute to volume with diluent. Filter through Nylon or PVDF filter 0.2 µm after discarding first 8.0 mL of filtrate

System suitability was performed by injecting resolution solution and determining resolution between closely eluting peak of Pramipexole, Pramipexole A, Pramipexole B, Pramipexole D and Pramipexole E. The TF and TP were checked in Standard solution. The standard solution for Pramipexole was prepared twice and injected. The parameters such as similarity factor, and RSD of RT , Area of Pramipexole was determined. Results for resolution system suitability are presented in **Table 49 and Table 50.**

Table 49: Result of System suitability-Accuracy (Related substances):

Analyte	Resolution
Pramipexole Imp A	N.A.
Pramipexole Imp B	2.1
Pramipexole	2.4
Pramipexole Imp E	7.1
Pramipexole Imp D	4.4

Table 50: Result of System suitability- Accuracy (Related substances):

Analyte	% RSD		T.P.	T.F.	Sim. Fac
	RT	Area			
Pramipexole Imp A	0.8	1.4	8902	1.2	0.96
Pramipexole Imp B	0.5	2.2	10783	1.2	0.98
Pramipexole	0.4	1.1	8673	1.1	0.99
Pramipexole Imp E	0.4	1.5	13211	1.0	0.99
Pramipexole Imp D	0.3	1.7	11783	1.1	0.96

The acceptance criterion for similarity factor was ≥0.98 ≤1.02, tailing factor for the peak due to Pramipexole, Pramipexole A, Pramipexole B, Pramipexole D and Pramipexole E. was NMT 2.0, theoretical plates was not less than 2000 and % RSD of RT , Area of Standard solution of first replicate solution was not more than 1.0% and 5.0% respectively.

The acceptance criteria were met.

3.5.7 Procedure:

The working Standard Solution is injected separately 5 times and sample solutions (50% solution, 100% solution & 150% solution) is also injected separately 3 times (about 5µl) respectively.

The % Recovery of Pramipexole is determined by the following Expression.

$$\text{Amount Found} = \frac{\text{Sample area}}{\text{Std Area}} \times \frac{\text{Std wt}}{\text{Std dilution}} \times 1000 \times \frac{\text{Purity}}{100}$$

$$\text{Amount Added} = \frac{\text{Std wt}}{\text{Std dilution}} \times \frac{\text{Purity}}{100} \times 1000$$

The % Recovery is determined by the following Expression.

$$\% \text{ Recovery} = \frac{\text{Amount Found}}{\text{Amount Added}} \times 100$$

The accuracy study of the analytes was determined for the following levels, 50%, 100% and 150% of the specified limit. The Analyte showed the recovery between 98% to 102 % concluding the accuracy of the method.

Table No. 51: % Recovery of Pramipexole known impurities Accuracy at LOQ Levels

% Level w.r.t. Limit Level	Preparation No.	% Recovery			
		Pramipexole Imp A	Pramipexole Imp D	Pramipexole Imp B	Pramipexole Imp E
LOQ	1	95.1	97	96.8	97.9
	2	96.4	96.2	97.5	98.9
	3	95.8	97.9	97.9	97.5
50	1	98.9	100.9	99.5	100.5

	2	99.4	100	99.4	100.9
	3	99.8	99.2	99.9	100.4
100	1	100.3	98.7	101.4	98.1
	2	100.8	98.3	100.9	99.8
	3	101.1	99.9	100.5	99.4
	4	99.9	100.5	101.9	100.9
	5	99.5	100.2	99.4	101.5
	6	100.9	99.1	99.8	100.2
150	1	101.7	100.9	100.1	98.5
	2	100.6	100.5	99.2	99.2
	3	101.3	99.5	98.2	100.8

Table No. 52: Concentration of Pramipexole known impurities Accuracy at LOQ Levels

% Level w.r.t. Limit Level	Preparation No.	Concentration (in μg per mL) of			
		Pramipexole Imp A	Pramipexole Imp D	Pramipexole Imp B	Pramipexole Imp E
LOQ	Mean	95.77	97.40	97.03	98.10
	SD	0.65	0.56	0.85	0.72
	RSD	0.68	0.57	0.88	0.74
50	Mean	99.37	99.60	100.03	100.60
	SD	0.45	0.26	0.85	0.26
	RSD	0.45	0.27	0.85	0.26
100	Mean	100.42	100.65	99.45	99.98
	SD	0.63	0.95	0.88	1.19
	RSD	0.62	0.94	0.89	1.19
150	Mean	101.20	99.17	100.30	99.50
	SD	0.56	0.95	0.72	1.18

| | RSD | 0.55 | 0.96 | 0.72 | 1.18 |

Table No. 53: Results of Accuracy study of Pramipexole Imp :

Sr. No.	Parameter	Results	Acceptance criteria
1	Individual % Recovery of Pramipexole Imp A	95.1% and 101.7%	85 % to 115 %
2	Individual % Recovery of Pramipexole Imp D	96.2% and 100.9%	85 % to 115 %
3	Individual % Recovery of Pramipexole Imp B	96.8% to 101.9%	85 % to 115 %
4	Individual % Recovery of Pramipexole Imp E	97.5% and 101.5%	85 % to 115 %
5	Mean % Recovery of Pramipexole Imp A	99.4%	90% to 110 %

6	Mean % Recovery of Pramipexole Imp D	99.3%	90% to 110 %
7	Mean % Recovery of Pramipexole Imp B	99.5%	90% to 110 %
8	Mean % Recovery of Pramipexole Imp E	99.6%	90% to 110 %

Conclusion: Related substance method is found accurate for as recovery results and individual recoveries of Pramipexole Imp A, Pramipexole (Imp D), Pramipexole Imp B, Pramipexole Imp E are well within the acceptance criteria.

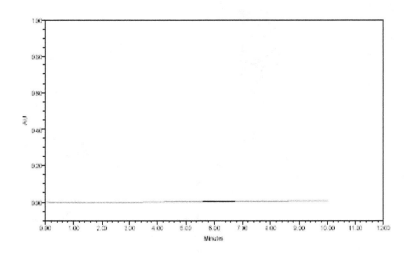

Figure 96: Chromatogram of Blank

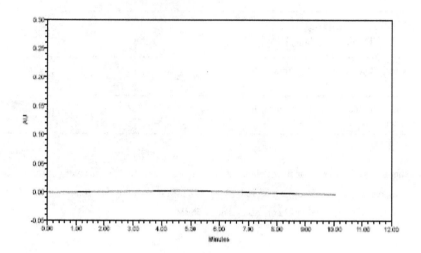

Figure 97: Chromatogram of placebo

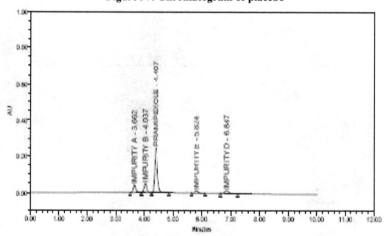

Figure 98 : Resolution chromatogram of Pramipexole and its related Impurities.

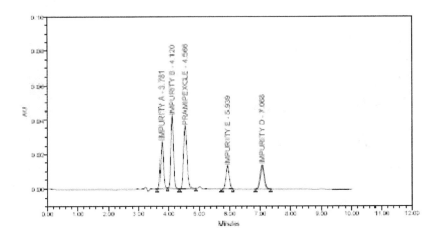

Figure 99: standard solution chromatogram

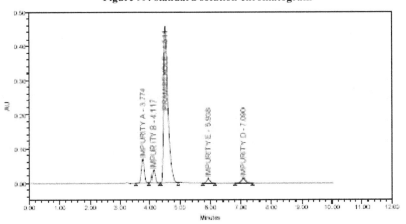

Figure 100: Accuracy 50% solution chromatogram

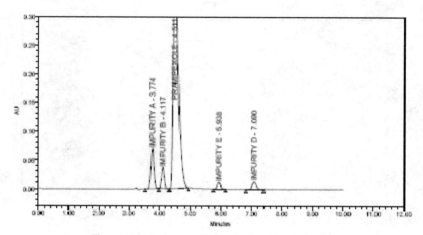

Figure 101: Accuracy 100% solution chromatogram

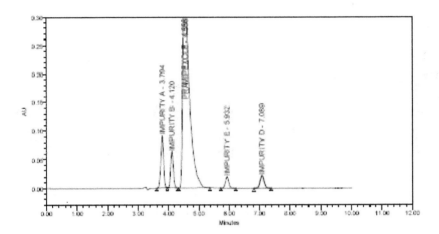

Figure 102: Accuracy 150% solution chromatogram

3.6 Precision :

System suitability solution, Standard solution, sample solution and Placebo solution preparation is as per specificity chapter.

3.6.1 Procedure:

The working standard solution is injected separately 6 times and six assay sample preparations are injected separately 1 times respectively (5µl). The chromatogram is recorded and measured the peak response.

System suitability was performed by injecting resolution solution and determining resolution between closely eluting peak of Pramipexole, Pramipexole A, Pramipexole B, Pramipexole D and Pramipexole E. The TF and TP were checked in Standard solution. The standard solution for Pramipexole was prepared twice and injected. The parameters such as similarity factor, and RSD of RT , Area of Pramipexole was determined. Results for resolution system suitability are presented in **Table 54 and Table 55.**.

Table 54: Result of System suitability-Precision (Related substances):

Analyte	Resolution
Pramipexole Imp A	N.A.
Pramipexole Imp B	2.1
Pramipexole	2.4
Pramipexole Imp E	7.1
Pramipexole Imp D	4.4

Table 55: Result of System suitability-Precision (Related substances):

Analyte	% RSD		T.P.	T.F.	Sim. Fac
	RT	Area			
Pramipexole Imp A	0.8	1.4	8902	1.2	0.96
Pramipexole Imp B	0.5	2.2	10783	1.2	0.98
Pramipexole	0.4	1.1	8673	1.1	0.99

| Pramipexole Imp E | 0.4 | 1.5 | 13211 | 1.0 | 0.99 |
| Pramipexole Imp D | 0.3 | 1.7 | 11783 | 1.1 | 0.96 |

The acceptance criterion for similarity factor was ≥0.98 ≤1.02, tailing factor for the peak due to Pramipexole, Pramipexole A, Pramipexole B, Pramipexole D and Pramipexole E. was NMT 2.0, theoretical plates was not less than 2000 and % RSD of RT , Area of Standard solution of first replicate solution was not more than 1.0% and 5.0% respectively. The acceptance criteria were met.

Samples of Pramipexole tablets does not shows any one of the Imp present in the sample hence in the accuracy and recovery chapter these known impurities were spiked for six replicates and the %RSD of that results are calculated for method precision chapter. The RSD for recovered Pramipexole A, Pramipexole B, Pramipexole D and Pramipexole E were within the limit of 5.00% confirming the precision of the method. **Table 56 and Table 59** represent the method precision results.

Table 56: Results of Precision Spiked of Pramipexole Imp A

		% Recovery	Mean	SD	%RSD
100%	1	100.3	100.42	0.63	0.62
	2	100.8			
	3	101.1			
	4	99.9			
	5	99.5			
	6	100.9			

Table 57: Results of Precision Spiked of Pramipexole Imp B

		% Recovery	Mean	SD	%RSD
100%	1	101.4	100.65	0.95	0.94
	2	100.9			
	3	100.5			
	4	101.9			
	5	99.4			
	6	99.8			

Table 58: Results of Precision Spiked of Pramipexole Imp D

		% Recovery	Mean	SD	%RSD
100%	1	98.7	99.45	0.88	0.89
	2	98.3			
	3	99.9			
	4	100.5			
	5	100.2			
	6	99.1			

Table 59: Results of Precision Spiked of Pramipexole Imp E

		% Recovery	Mean	SD	%RSD
100%	1	98.1	99.98	1.19	1.19
	2	99.8			
	3	99.4			
	4	100.9			
	5	101.5			
	6	100.2			

The relative standard deviation of six sample preparation Pramipexole A, Pramipexole B, Pramipexole D and Pramipexole E each, from the tablet solution was found to be within the acceptance criteria of not more than 5.00%. Thus, illustrating the HPLC analytical method for the % recovery of Impurities of Pramipexole tablets is precise.

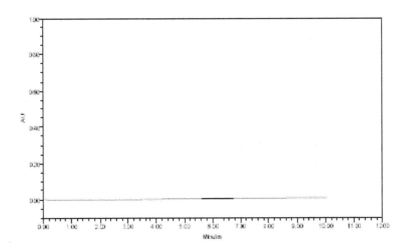

Figure 103: Chromatogram of Blank

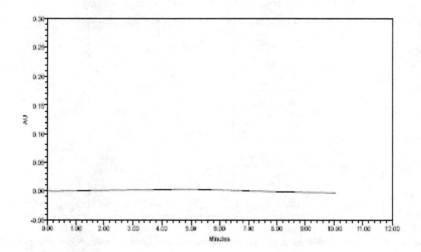

Figure 104: Chromatogram of placebo

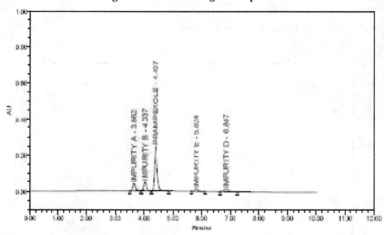

Figure 105 : Resolution chromatogram of Pramipexole and its related Impurities.

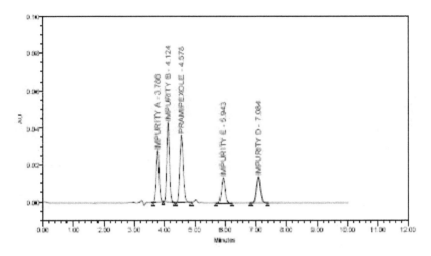

Figure 106: standard solution chromatogram

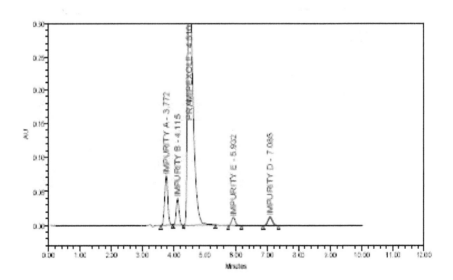

Figure 107: Precision sample solution chromatogram

3.7 Solution Stability:

Stability of solutions in Room temperature from 0th hr, 2nd hr, 4th hr, 6th hr, 8th hr, 12th hr, 16th hr, 20th hr and then up to 24th hrs. System suitability solution, Standard solution, sample solution and Placebo solution preparation is as per specificity chapter.

3.7.1 Procedure:

MP and fresh standard was prepared for system suitability. The working standard solution (freshly prepared) and the Related sample solution (freshly prepared) is injected separately 6 and 1 replicates respectively and this solution was kept in room temperature from 0hrs to 24th hr.

System suitability was performed by injecting resolution solution and determining resolution between closely eluting peak of Pramipexole, Pramipexole A, Pramipexole B, Pramipexole D and Pramipexole E. The TF and TP were checked in Standard solution. The standard solution for Pramipexole was prepared twice and injected. The parameters such as similarity factor, and RSD of RT, Area of Pramipexole was determined. Results for resolution system suitability are presented in **Table 60 and Table 61**.

Table 60: Result of System suitability-Solution Stability (Related substances):

Analyte	Resolution
Pramipexole Imp A	N.A.
Pramipexole Imp B	2.0
Pramipexole	2.4
Pramipexole Imp E	6.7
Pramipexole Imp D	4.8

Table 61: Result of System suitability-Solution Stability (Related substances):

Analyte	% RSD		T.P.	T.F.	Sim. Fac
	RT	Area			
Pramipexole Imp A	0.9	3.4	9105	1.1	0.99
Pramipexole Imp B	0.9	3.0	11408	0.9	1.04
Pramipexole	0.3	2.2	8143	1.6	1.01
Pramipexole Imp E	0.4	2.0	10837	1.3	1.01

| Pramipexole Imp D | 0.3 | 2.5 | 11902 | 1.3 | 1.02 |

The acceptance criterion for similarity factor was ≥0.98 ≤1.02, tailing factor for the peak due to Pramipexole, Pramipexole A, Pramipexole B, Pramipexole D and Pramipexole E. was NMT 2.0, theoretical plates was not less than 2000 and % RSD of RT , Area of Standard solution of first replicate solution was not more than 1.0% and 5.0% respectively. The acceptance criteria were met.

The stability of sample solutions were determined at time periods representative for storage. The stability of sample solution was confirmed on the sample by comparing the values of 0th sample at different time interval to its initial value. The % difference for recovery of Imp should not be more than 5.00%. The data for sample solution stability in **Table 62**.

Table 62: Sample solution stability of Pramipexole A, Pramipexole B, Pramipexole D and Pramipexole E:

Sample Hr	% Recovery of Imp A	% Recovery of Imp B	% Recovery of Imp D	% Recovery of Imp E
0th hr	98.2	100.1	101.9	99.5
2nd hr	99.1	100.9	100.2	99
4th hr	98.5	101.8	101.6	101.8
8th hr	98.9	101.3	101.4	101.1
12th hr	100.7	99.2	99.2	100.2
16th hr	100.2	98.1	98.4	101.6
20th hr	101.8	98.7	100.6	101.7
24th hr	101.3	99.6	101.7	99.5
Mean	**99.84**	**99.96**	**100.63**	**100.55**
SD	1.35	1.30	1.28	1.14
%RSD	**1.35**	**1.30**	**1.27**	**1.13**

The standard was found stable till 24 hrs since the RSD of the % recovery of Pramipexole A, Pramipexole B, Pramipexole D and Pramipexole E is not more than 5.00 %. Which shows sample solution is also stable for 24 hrs.

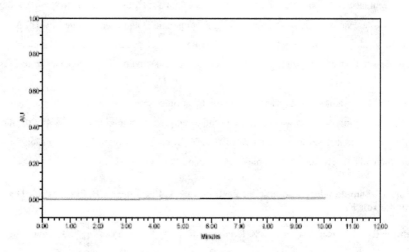

Figure 108: Chromatogram of Blank

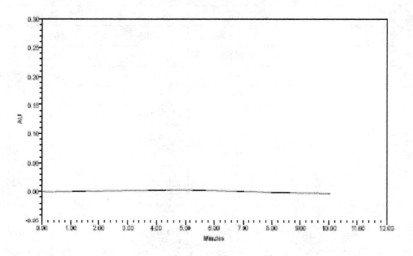

Figure 109: Chromatogram of placebo

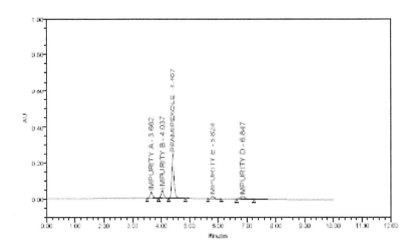

Figure 110 : Resolution chromatogram of Pramipexole and its related Impurities.

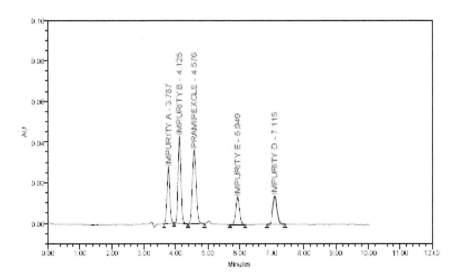

Figure 111: standard solution chromatogram

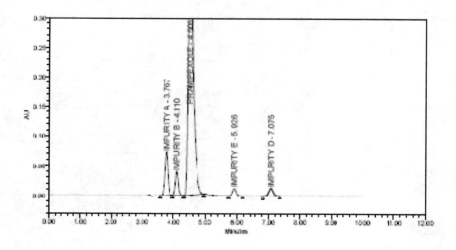

Figure 112: Solution Stability sample solution chromatogram

3.8 Intermediate Precision or Ruggedness:

System suitability solution, Standard solution, sample solution and Placebo solution preparation is as per specificity chapter.

3.8.1 Procedure:

The working standard solution is injected separately 6 times and six assay sample preparations are injected separately 1 times respectively (5µl). The chromatogram is recorded and measured the peak response.

System suitability was performed by injecting resolution solution and determining resolution between closely eluting peak of Pramipexole, Pramipexole A, Pramipexole B, Pramipexole D and Pramipexole E. The TF and TP were checked in Standard solution. The standard solution for Pramipexole was prepared twice and injected. The parameters such as similarity factor, and RSD of RT, Area of Pramipexole was determined. Results for resolution system suitability are presented in **Table 63 and Table 64**.

Table 63: Result of System suitability-Intermediate Precision (Related substances):

Analyte	Resolution
Pramipexole Imp A	N.A.
Pramipexole Imp B	2.2
Pramipexole	2.3
Pramipexole Imp E	6.6
Pramipexole Imp D	4.5

Table 64: Result of System suitability-Intermediate precision (Related substances):

Analyte	% RSD		T.P.	T.F.	Sim. Fac
	RT	Area			
Pramipexole Imp A	0.6	2.2	9474	1.0	1.02
Pramipexole Imp B	0.6	3.1	12076	1.0	1.02
Pramipexole	0.8	2.0	8829	1.8	1.02
Pramipexole Imp E	0.5	2.4	11760	1.0	1.03

| Pramipexole Imp D | 0.6 | 2.9 | 12609 | 1.4 | 1.00 |

The acceptance criterion for similarity factor was ≥0.98 ≤1.02, tailing factor for the peak due to Pramipexole, Pramipexole A, Pramipexole B, Pramipexole D and Pramipexole E. was NMT 2.0, theoretical plates was not less than 2000 and % RSD of RT , Area of Standard solution of first replicate solution was not more than 1.0% and 5.0% respectively. The acceptance criteria were met.

Analysis done on six sample prepared from same batch by a another chemist freshly prepared. The RSD of % recovery of known Imp was evaluated and was found in limit of NMT 5.00%. The observed results are in **Table 65**.

Table 65: Results of Intermediate Precision Study

Instrument	Waters – Empower		Waters – Empower	
Analyst	A		B	
	% Recovery of Pramipexole Imp A		% Recovery of Pramipexole Imp B	
Sr. No.	(A)	(B)	(A)	(B)
1.	100.3	98.1	101.4	99.2
2.	100.8	99.3	100.9	99.1
3.	101.1	98.4	100.5	100.8
4.	99.9	99.9	101.9	100.1
5	99.5	100.3	99.4	98.5
6	100.9	101.2	99.8	99.7
Mean	99.98		100.11	
SD	1.01		1.01	
%RSD	1.01		1.01	

Table 66: Results of Intermediate Precision Study

Instrument	Waters – Empower		Waters – Empower	
Analyst	A		B	
	% Recovery of Pramipexole Imp D		% Recovery of Pramipexole Imp E	
Sr. No.	(A)	(B)	(A)	(B)
1.	98.7	101.8	98.1	100.8
2.	98.3	100.5	99.8	100.4
3.	99.9	100.6	99.4	100.9
4.	100.5	99.7	100.9	99.1
5	100.2	98.2	101.5	98.2
6	99.1	98.9	100.2	101.9
Mean	99.70		100.10	
SD	1.09		1.22	
%RSD	1.09		1.21	

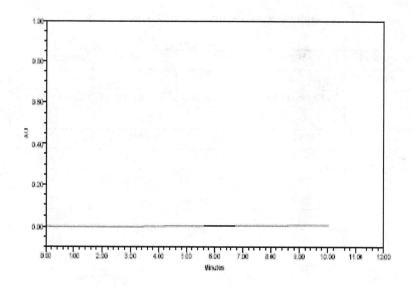

Figure 113: Chromatogram of Blank

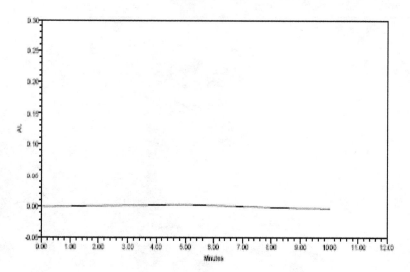

Figure 114: Chromatogram of placebo

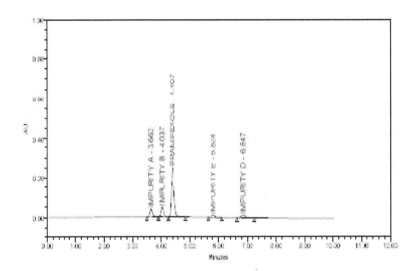

Figure 115 : Resolution chromatogram of Pramipexole and its related Impurities.

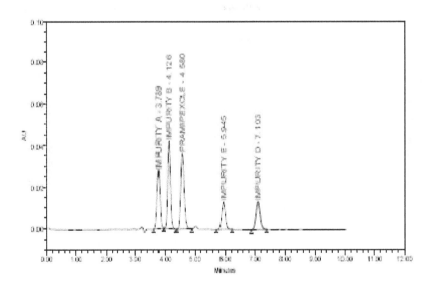

Figure 116: standard solution chromatogram

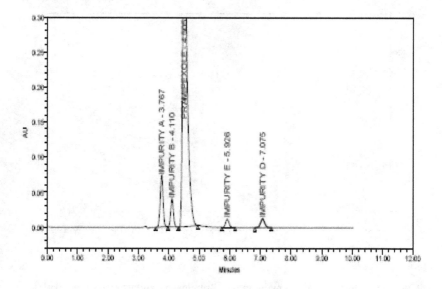

Figure 117: Intermediate Precision sample solution chromatogram

3.9 Robustness:

In this chapters mainly changed one parameter deliberately in chromatographic conditions carried out (flow rate and column oven temperature in the variants).

3.9.1 Robustness-I: (Flow rate changed by –0.1ml)

System suitability solution, Standard solution, sample solution and Placebo solution preparation is as per specificity chapter.

3.9.2 Procedure:

The working standard solution is injected separately 6 times and three assay sample preparations are injected separately 1 times (5µl). The chromatogram is recorded and measured the peak response.

System suitability was performed by injecting resolution solution and determining resolution between closely eluting peak of Pramipexole, Pramipexole A, Pramipexole B, Pramipexole D and Pramipexole E. The TF and TP were checked in Standard solution. The standard solution for Pramipexole was prepared twice and injected. The parameters such as similarity factor, and RSD of RT , Area of Pramipexole was determined. Results for resolution system suitability are presented in **Table 67 and Table 68.**

Table 67: Result of System suitability-Robustness I (Related substances):

Analyte	Resolution
Pramipexole Imp A	N.A.
Pramipexole Imp B	2.2
Pramipexole	2.3
Pramipexole Imp E	6.5
Pramipexole Imp D	4.7

Table 68: Result of System suitability-Robustness I (Related substances):

Analyte	% RSD		T.P.	T.F.	Sim. Fac
	RT	Area			
Pramipexole Imp A	0.5	3.1	8849	1.2	1.03
Pramipexole Imp B	0.5	2.7	11006	1.1	1.01
Pramipexole	0.2	1.7	8756	1.4	1.00
Pramipexole Imp E	0.6	1.9	12533	1.2	1.00

| Pramipexole Imp D | 0.2 | 2.5 | 10691 | 1.6 | 1.04 |

The acceptance criterion for similarity factor was ≥0.98 ≤1.02, tailing factor for the peak due to Pramipexole, Pramipexole A, Pramipexole B, Pramipexole D and Pramipexole E. was NMT 2.0, theoretical plates was not less than 2000 and % RSD of RT , Area of Standard solution of first replicate solution was not more than 1.0% and 5.0% respectively. The acceptance criteria were met.

The resolution does not show significant changes in robustness. This shows minor change of the parameters does not affect the chromatographic separation. It was analyzed on three different sample solutions from same analytical batch. The % RSD was evaluated and found within the limit of NMT 5.00%. The results were presented in **Table 69**.

Table 69: Robustness-1 Results of % recovery of Pramipexole

Sample No.	% Recovery of Pramipexole Imp A	% Recovery of Pramipexole Imp B	% Recovery of Pramipexole Imp D	% Recovery of Pramipexole Imp E
1	101.3	99.2	101.8	100.8
2	100.8	99.1	100.5	100.4
3	99.3	100.8	100.6	100.9
Mean	**100.47**	**99.70**	**100.97**	**100.70**
SD	1.04	0.95	0.72	0.26
%RSD	**1.04**	**0.96**	**0.72**	**0.26**

The cumulative % RSD was evaluated with the method precision samples and was within the acceptance criterion of NMT 5.00%. The results were presented in **Table 70**.

Table 70: Robustness-I Results of cumulative RSD % recovery of Pramipexole

Sample No.	% Recovery of Pramipexole Imp A	% Recovery of Pramipexole Imp B	% Recovery of Pramipexole Imp E	% Recovery of Pramipexole Imp D
1	101.3	99.2	101.8	100.8
2	100.8	99.1	100.5	100.4
3	99.3	100.8	100.6	100.9
4	98.1	99.2	101.8	100.8
5	99.3	99.1	100.5	100.4
6	98.4	100.8	100.6	100.9
7	99.9	100.1	99.7	99.1
8	100.3	98.5	98.2	98.2
9	101.2	99.7	98.9	101.9
Mean	**99.84**	**99.61**	**100.29**	**100.38**
SD	1.16	0.80	1.20	1.10
%RSD	**1.16**	**0.81**	**1.19**	**1.09**

3.9.3 Robustness -II:(Flow rate changed by +0.1ml)
(Resolution standard solution, Standard solution, Sample solution (three preparations and placebo solution prepared as robustness-I chapter)

3.9.4 Procedure:

The working standard solution is injected separately 6 times and three assay sample preparations are injected separately 1 times (5 µl). The chromatogram is recorded and measured the peak response.

System suitability was performed by injecting resolution solution and determining resolution between closely eluting peak of Pramipexole, Pramipexole A, Pramipexole B, Pramipexole D and Pramipexole E. The TF and TP were checked in Standard solution. The standard solution for Pramipexole was prepared twice and injected. The parameters such as similarity factor, and RSD of RT , Area of Pramipexole was determined. Results for resolution system suitability are presented in **Table 71 and Table 72**.

Table 71: Result of System suitability-Robustness II (Related substances):

Analyte	Resolution
Pramipexole Imp A	N.A.
Pramipexole Imp B	2.1
Pramipexole	2.2
Pramipexole Imp E	6.8
Pramipexole Imp D	4.6

Table 72: Result of System suitability-Robustness II (Related substances):

Analyte	% RSD		T.P.	T.F.	Sim. Fac
	RT	Area			
Pramipexole Imp A	0.6	2.3	8709	1.0	1.02
Pramipexole Imp B	0.6	1.1	9960	1.2	1.00
Pramipexole	0.4	1.4	9056	1.7	1.03
Pramipexole Imp E	0.4	1.2	11480	1.4	1.02

| Pramipexole Imp D | 0.5 | 1.2 | 11052 | 1.3 | 1.00 |

The acceptance criterion for similarity factor was ≥0.98 ≤1.02, tailing factor for the peak due to Pramipexole, Pramipexole A, Pramipexole B, Pramipexole D and Pramipexole E. was NMT 2.0, theoretical plates was not less than 2000 and % RSD of RT , Area of Standard solution of first replicate solution was not more than 1.0% and 5.0% respectively. The acceptance criteria were met.

The resolution does not show significant changes in robustness. This shows minor change of the parameters does not affect the chromatographic separation. It was analyzed on three different sample solutions from same analytical batch. The % RSD was evaluated and found within the limit of NMT 5.00%. The results were presented in **Table 73**.

Table 73: Robustness-II Results of % recovery of Pramipexole

Sample No.	% Recovery of Pramipexole Imp A	% Recovery of Pramipexole Imp B	% Recovery of Pramipexole Imp E	% Recovery of Pramipexole Imp D
1	98.2	101.2	101.7	101.9
2	99.4	100.7	100.6	100.4
3	100.2	99.5	99.2	99.6
Mean	**99.27**	**100.47**	**100.50**	**100.63**
SD	1.01	0.87	1.25	1.17
%RSD	**1.01**	**0.87**	**1.25**	**1.16**

The cumulative % RSD was evaluated with the method precision samples and was within the acceptance criterion of NMT 5.00% .The results were presented in **Table 74**.

Table 74: Robustness-II Results of cumulative RSD % recovery of Pramipexole

Sample No.	% Recovery of Pramipexole Imp A	% Recovery of Pramipexole Imp B	% Recovery of Pramipexole Imp E	% Recovery of Pramipexole Imp D
1	98.2	101.2	101.7	101.9
2	99.4	100.7	100.6	100.4
3	100.2	99.5	99.2	99.6
4	98.1	99.2	101.8	100.8
5	99.3	99.1	100.5	100.4
6	98.4	100.8	100.6	100.9
7	99.9	100.1	99.7	99.1
8	100.3	98.5	98.2	98.2
9	101.2	99.7	98.9	101.9
Mean	**99.44**	**99.87**	**100.13**	**100.36**
SD	1.06	0.90	1.23	1.23
%RSD	**1.07**	**0.90**	**1.23**	**1.22**

3.9.5 Robustness -III: (Column oven Temp. changed by +2)
(Resolution standard solution, Standard solution, Sample solution (three preparations and placebo solution prepared as robustness-I chapter)

3.9.6 Procedure:
The working standard solution is injected separately 6 times and three assay sample preparations are injected separately 1 times (5 µl). The chromatogram is recorded and measured the peak response.

System suitability was performed by injecting resolution solution and determining resolution between closely eluting peak of Pramipexole, Pramipexole A, Pramipexole B, Pramipexole D and Pramipexole E. The TF and TP were checked in Standard solution. The standard solution for Pramipexole was prepared twice and injected. The parameters such as similarity factor, and RSD of RT, Area of Pramipexole was determined. Results for resolution system suitability are presented in **Table 75 and Table 76.**.

Table 75: Result of System suitability-Robustness III (Related substances):

Analyte	Resolution
Pramipexole Imp A	N.A.
Pramipexole Imp B	2.2
Pramipexole	2.3
Pramipexole Imp E	6.6
Pramipexole Imp D	4.5

Table 76: Result of System suitability-Robustness III (Related substances):

Analyte	% RSD		T.P.	T.F.	Sim. Fac
	RT	Area			
Pramipexole Imp A	0.4	1.9	8995	1.1	1.01
Pramipexole Imp B	0.4	1.0	9827	1.3	0.97
Pramipexole	0.3	1.0	9320	1.9	1.00
Pramipexole Imp E	0.5	1.5	12009	1.1	0.99

| Pramipexole Imp D | 0.7 | 1.8 | 13002 | 1.2 | 0.97 |

The acceptance criterion for similarity factor was ≥0.98 ≤1.02, tailing factor for the peak due to Pramipexole, Pramipexole A, Pramipexole B, Pramipexole D and Pramipexole E. was NMT 2.0, theoretical plates was not less than 2000 and % RSD of RT , Area of Standard solution of first replicate solution was not more than 1.0% and 5.0% respectively. The acceptance criteria were met.

The resolution does not show significant changes in robustness. This shows minor change of the parameters does not affect the chromatographic separation. It was analyzed on three different sample solutions from same analytical batch. The % RSD was evaluated and found within the limit of NMT 5.00%. The results were presented in **Table 77**.

Table 77: Robustness-III Results of % recovery of Pramipexole

Sample No.	% Recovery of Pramipexole Imp A	% Recovery of Pramipexole Imp B	% Recovery of Pramipexole Imp E	% Recovery of Pramipexole Imp D
1	101.2	99.8	99.7	100.8
2	101.9	98.4	101.3	101.6
3	100.5	101.5	100.4	99
Mean	**101.20**	**99.90**	**100.47**	**100.47**
SD	0.70	1.55	0.80	1.33
%RSD	**0.69**	**1.55**	**0.80**	**1.33**

The cumulative % RSD was evaluated with the method precision samples and was within the acceptance criterion of NMT 5.00%. The results were presented in **Table 78**.

Table 78: Robustness-III Results of cumulative RSD % recovery of Pramipexole

Sample No.	% Recovery of Pramipexole Imp A	% Recovery of Pramipexole Imp B	% Recovery of Pramipexole Imp E	% Recovery of Pramipexole Imp D
1	101.2	99.8	99.7	100.8
2	101.9	98.4	101.3	101.6
3	100.5	101.5	100.4	99
4	98.1	99.2	101.8	100.8
5	99.3	99.1	100.5	100.4
6	98.4	100.8	100.6	100.9
7	99.9	100.1	99.7	99.1
8	100.3	98.5	98.2	98.2
9	101.2	99.7	98.9	101.9
Mean	**100.09**	**99.68**	**100.12**	**100.30**
SD	1.30	1.02	1.13	1.26
%RSD	**1.29**	**1.03**	**1.13**	**1.25**

3.9.7 Robustness -IV :(Column oven Temp. Changed by -2)
(Resolution standard solution, Standard solution, Sample solution (three preparations and placebo solution prepared as robustness-I chapter)

3.9.8 Procedure:
The working standard solution is injected separately 5 times and three assay sample preparations are injected separately 1 times (about 5 µl). The chromatogram is recorded and measured the peak response.

System suitability was performed by injecting resolution solution and determining resolution between closely eluting peak of Pramipexole, Pramipexole A, Pramipexole B, Pramipexole D and Pramipexole E. The TF and TP were checked in Standard solution. The standard solution for Pramipexole was prepared twice and injected. The parameters such as similarity factor, and RSD of RT , Area of Pramipexole was determined. Results for resolution system suitability are presented in **Table 79 and Table 80**.

Table 79: Result of System suitability-Robustness IV (Related substances):

Analyte	Resolution
Pramipexole Imp A	N.A.
Pramipexole Imp B	2.1
Pramipexole	2.2
Pramipexole Imp E	6.7
Pramipexole Imp D	4.7

Table 80: Result of System suitability-Robustness IV (Related substances):

Analyte	% RSD		T.P.	T.F.	Sim. Fac
	RT	Area			
Pramipexole Imp A	0.3	1.7	9002	1.4	0.98
Pramipexole Imp B	0.3	2.0	11049	1.1	1.00
Pramipexole	0.1	1.5	9810	1.3	0.98
Pramipexole Imp E	0.2	1.2	11689	1.3	1.01

| Pramipexole Imp D | 0.4 | 1.0 | 12604 | 1.0 | 1.02 |

The acceptance criterion for similarity factor was ≥0.98 ≤1.02, tailing factor for the peak due to Pramipexole, Pramipexole A, Pramipexole B, Pramipexole D and Pramipexole E. was NMT 2.0, theoretical plates was not less than 2000 and % RSD of RT , Area of Standard solution of first replicate solution was not more than 1.0% and 5.0% respectively. The acceptance criteria were met.

The resolution does not show significant changes in robustness. This shows minor change of the parameters does not affect the chromatographic separation. It was analyzed on three different sample solutions from same analytical batch. The % RSD was evaluated and found within the limit of NMT 5.00%. The results were presented in **Table 81**.

Table 81: Robustness-IV Results of % recovery of Pramipexole

Sample No.	% Recovery of Pramipexole Imp A	% Recovery of Pramipexole Imp B	% Recovery of Pramipexole Imp E	% Recovery of Pramipexole Imp D
1	100.8	101.7	100.4	101.6
2	101.5	100.4	101.8	99.8
3	99.7	100.9	100	100.7
Mean	**100.67**	**101.00**	**100.73**	**100.70**
SD	0.91	0.66	0.95	0.90
%RSD	**0.90**	**0.65**	**0.94**	**0.89**

The cumulative % RSD was evaluated with the method precision samples and was within the acceptance criterion of NMT 5.00%. The results were presented in **Table 82**.

Table 82: Robustness-IV Results of cumulative RSD % recovery of Pramipexole

Sample No.	% Recovery of Pramipexole Imp A	% Recovery of Pramipexole Imp B	% Recovery of Pramipexole Imp E	% Recovery of Pramipexole Imp D
1	100.8	101.7	100.4	101.6
2	101.5	100.4	101.8	99.8
3	99.7	100.9	100	100.7
4	98.1	99.2	101.8	100.8
5	99.3	99.1	100.5	100.4
6	98.4	100.8	100.6	100.9
7	99.9	100.1	99.7	99.1
8	100.3	98.5	98.2	98.2
9	101.2	99.7	98.9	101.9
Mean	**99.91**	**100.04**	**100.21**	**100.38**
SD	1.18	1.02	1.19	1.18
%RSD	**1.18**	**1.02**	**1.19**	**1.17**

Chapter 4

Summary and Conclusion

4.1 Summary and Conclusion for Assay :

4.1.1 Resolution System suitability for Assay:

The suitability was evaluated by injecting resolution solution. Resolution between Pramipexole Imp A, Pramipexole Imp B, Pramipexole, Pramipexole Imp E and Pramipexole Imp D was evaluated. Results for resolution system suitability are as follows:

Table-83: Resolution System suitability for Assay:

Experiment	Pramipexole Imp A and Pramipexole Imp B	Pramipexole Imp B and Pramipexole	Pramipexole and Pramipexole Imp E	Pramipexole Imp E and Pramipexole Imp D
Specificity	2.1	2.3	6.7	4.6
Precision	2.0	2.4	6.9	4.5
Accuracy	2.2	2.5	7.0	4.4
Intermediate Precision	2.1	2.4	6.5	4.6
Solution Stability	2.2	2.3	6.6	4.7
Robustness-I	2.3	2.2	6.7	4.6
Robustness-II	2.2	2.3	6.8	4.7
Robustness-III	2.1	2.4	6.5	4.8
Robustness-IV	2.0	2.3	6.6	4.6

The acceptance criterion for resolution was not less than 2.0 between any two consecutive peaks.

The TF and TP were checked in Standard solution for the first replicate of standard solution A. The standard solution for Pramipexole was prepared twice and injected. The parameters such as similarity factor, and RSD of RT , Area of Pramipexole was determined. Results for system suitability are as follows:

Table-84: System Suitability for Pramipexole:

Experiment		% RSD		T.P	T.F.	Sim. Fac.
		RT	Area			
Specificity		0.0	0.4	12385	1.2	1.00
Linearity	50%	0.3	0.8	NA	NA	NA
	150%	0.2	0.2	NA	NA	NA
Accuracy		0.4	0.3	15467	1.3	0.99
Precision		0.5	0.4	14256	1.2	0.99
Intermediate Precision		0.2	0.9	13024	0.9	0.99
Solution Stability		0.3	0.7	12890	1.0	0.98
Robustness-I		0.5	1.4	13856	1.2	1.00
Robustness-II		0.2	0.4	13408	1.1	1.01
Robustness-III		0.4	1.2	12609	1.3	1.00
Robustness-IV		0.2	0.3	12589	1.2	0.99

The acceptance criterion for similarity factor was ≥0.98 ≤1.02, tailing factor for the peaks due to Pramipexole was NMT 2.0, theoretical plates was not less than 2000 and % RSD of RT , Area of Pramipexole Standard solution of first replicate solution was not more than 1.0% and 2.0% respectively. The acceptance criteria were met.

Blank (diluent) was injected to check the interference to Pramipexole Imp A, Pramipexole Imp B, Pramipexole, Pramipexole Imp E and Pramipexole Imp D. No interfernce to the peaks of Pramipexole Imp A, Pramipexole Imp B, Pramipexole, Pramipexole Imp E and Pramipexole Imp D.

4.1.2 Linearity:

Five Solutions containing concentrations of Pramipexole in the range of 50% to 150% of the working level (i.e. 102 µg/ml to 306 µg/ml) were injected into the chromatograph. The peak area responses were found to be linear with respect to the concentration. The regression coefficient (r^2) for Pramipexole was determined as 1.000 , the % Y-intercept for Pramipexole was -0.13% , the Response factor for Pramipexole was 1.1% .

The criteria for regression coefficient was not less than 0.999, % Y-intercept was not more than ± 2.0% The % RSD of responses factor of Pramipexole peak was not more than 3.00%. The regression coefficient, the %Y-intercept, RSD of area and RT and % RSD of responses factor were within the acceptance range.

4.1.3 Precision -Repeatability:
The sample solution was prepared six times. The content of Pramipexole for each preparation was determined. The % RSD for the content of Pramipexole of all six samples was 0.94%.
This met the acceptance criteria of not more than 2.00%.

4.1.4 Accuracy & Recovery:
The sample solution was prepared after spiking of known concentration in placebo solution three times at each level (i.e. 50%, 100% and 150%). The recovery of Pramipexole for each preparation was determined. The % RSD for the Pramipexole recovery of all three samples was 0.30% for 50%, 0.26% for 100% and 1.07% for 150% respectively.
This met the acceptance criteria of not more than 2.00%.

4.1.5 Precision - Intermediate precision:
System suitability and the Precision of Samples were carried out by second analyst on same system and on different day.
The sample solution was prepared six times. The content of Pramipexole for each preparation was determined. The % RSD for the Pramipexole content of all six samples was 1.10%. This met the acceptance criteria of not more than 2.00%.
The % variation of the average content of Pramipexole from precision- repeatability study and precision-intermediate precision study was 0.94%. This met the acceptance criteria of not more than 3.0 %.

4.1.6 Solution Stability:
The content of Pramipexole was determined initially and then at the interval up to the 24 hrs. System suitability of standard solution injected at varying time intervals are:
Table-85: For Pramipexole:

Interval	Tailing factor	Theoretical plates	% RSD of Retention time	% RSD of Area
Initial	1.0	12890	0.3	0.7
24 hrs	1.3	12589	0.1	1.4

The acceptance criteria were met.
The % relative difference for the content of Pramipexole calculated at 24th hr time interval is as follows in **Table no.86:**

VALIDATION TEST	OBSERVATION	RESULT	ACCEPTANCE CRITERIA
Solution stability study	% Relative difference for Pramipexole		
	24 hrs	0.5 %	NMT 3.0%

No additional peaks were observed for samples after the specified interval when compared with the initial run. Hence the sample solution was found to be stable up to 24 hours at laboratory conditions.

4.1.7 Robustness Exp-I :

Robustness Exp I was carried out by changing the flow rate of MP by – 0.1ml. The flow rate of MP changed from 1.0 ml to 0.9 ml per minute. Specificity, System suitability and the Precision of Samples were performed.

The sample solution was prepared three times. The content of Pramipexole for each preparation was determined. The % RSD for the Pramipexole content of all three samples was 0.57% . This met the acceptance criteria of not more than 2.00%.

The % Relative difference of the average content of Pramipexole in precision-repeatability study and Robustness-I study was 1.64% . This met the acceptance criteria of not more than 3.0%.

4.1.8 Robustness Exp-II:

Robustness Exp I was carried out by changing the flow rate of MP by +0.1ml. The flow rate of MP changed from 1.0 ml to 1.1 ml per minute. Specificity, System suitability and the Precision of Samples were performed.

The sample solution was prepared three times. The content of Pramipexole for each preparation was determined. The % RSD for the Pramipexole content of all three samples was 0.65%. This met the acceptance criteria of not more than 2.00%.

The % Relative difference of the average content of Pramipexole in precision-repeatability study and Robustness-II study was 0.98%. This met the acceptance criteria of not more than 3.0%.

4.1.9 Robustness Exp-III:

Robustness Exp-III was carried out by changing the column oven temperature by -2 °C. The column oven temperature changed from 40 °C to 38 °C. Specificity, System suitability and the Precision of Samples were performed.

The sample solution was prepared three times. The content of Pramipexole for each preparation was determined. The % RSD for the Pramipexole content of all three samples was 0.93%. This met the acceptance criteria of not more than 2.00%.

The % Relative difference of the average content of Pramipexole in precision-repeatability study and Robustness-III study was 1.49%. This met the acceptance criteria of not more than 3.0%.

4.1.10 Robustness Exp-IV:

Robustness Exp-IV was carried out by changing the column oven temperature by +2 °C. The column oven temperature changed from 40 °C to 42 °C. Specificity, System suitability and the Precision of Samples were performed.

The sample solution was prepared three times. The content of Pramipexole for each preparation was determined. The % RSD for the Pramipexole content of all three samples was 0.61%. This met the acceptance criteria of not more than 2.00%.

The % Relative difference of the average content of Pramipexole in precision-repeatability study and Robustness-IV study was 1.38%. This met the acceptance criteria of not more than 3.0%.

4.2 Related substances:
4.2.1 Resolution System suitability for Related substances:

The suitability was evaluated by injecting resolution solution. Resolution between Pramipexole Imp A, Pramipexole Imp B, Pramipexole, Pramipexole Imp E and Pramipexole Imp D was evaluated.

Table-87: Resolution System suitability for Related Substances :

Experiment	Pramipexole Imp A and Pramipexole Imp B	Pramipexole Imp B and Pramipexole	Pramipexole and Pramipexole Imp E	Pramipexole Imp E and Pramipexole Imp D
Specificity	2.2	2.5	6.8	4.6
Precision	2.1	2.4	7.1	4.4
Accuracy	2.1	2.4	7.1	4.4
Intermediate Precision	2.2	2.3	6.6	4.5
Solution Stability	2.0	2.4	6.7	4.8
Robustness-I	2.2	2.3	6.5	4.7

Robustness-II	2.1	2.2	6.8	4.6
Robustness-III	2.2	2.3	6.6	4.5
Robustness-IV	2.1	2.2	6.7	4.7

The acceptance criterion for resolution was not less than 2.0 between any two consecutive peaks. The TF and TP were checked in Standard solution for the first replicate of standard solution A. The standard solution for Pramipexole Imp A, Pramipexole Imp B, Pramipexole, Pramipexole Imp E and Pramipexole Imp D was prepared twice and injected. The parameters such as similarity factor, and RSD of RT , Area of Pramipexole Imp A, Pramipexole Imp B, Pramipexole, Pramipexole Imp E and Pramipexole Imp D were determined. Results for resolution system suitability are as follows:

Table-88: System Suitability for Pramipexole Imp A :

Experiment		% RSD		T.P	T.F.	Sim. Fac.
		RT	Area			
Specificity		0.8	1.5	8628	1.2	0.98
Linearity	LOQ	0.4	1.8	NA	NA	NA
	150%	0.3	0.4	NA	NA	NA
Accuracy		0.8	1.4	8902	1.2	0.96
Precision		0.8	1.4	8902	1.2	0.96
Intermediate Precision		0.6	2.2	9474	1.0	1.02
Solution Stability		0.9	3.4	9105	1.1	0.99
Robustness-I		0.5	3.1	8849	1.2	1.03
Robustness-II		0.6	2.3	8709	1.0	1.02
Robustness-III		0.4	1.9	8995	1.1	1.01
Robustness-IV		0.3	1.7	9002	1.4	0.98

Table-89: System Suitability for Pramipexole Imp B :

Experiment	% RSD	T.P	T.F.	Sim. Fac.

Experiment		% RSD		T.P	T.F.	Sim. Fac.
		RT	Area			
Specificity		0.6	1.8	10108	1.0	0.96
Linearity	LOQ	0.8	2.9	NA	NA	NA
	150%	0.3	0.5	NA	NA	NA
Accuracy		0.5	2.2	10783	1.2	0.98
Precision		0.5	2.5	10783	1.2	0.98
Intermediate Precision		0.6	3.1	12076	1.0	1.02
Solution Stability		0.9	3.0	11408	0.9	1.04
Robustness-I		0.5	2.7	11006	1.1	1.01
Robustness-II		0.6	1.1	9960	1.2	1.00
Robustness-III		0.4	1.0	9827	1.3	0.97
Robustness-IV		0.3	2.0	11049	1.1	1.00

Table-90: System Suitability for Pramipexole :

Experiment		% RSD		T.P	T.F.	Sim. Fac.
		RT	Area			
Specificity		0.7	1.1	8396	1.0	0.98
Linearity	LOQ	0.9	2.2	NA	NA	NA
	150%	0.2	0.2	NA	NA	NA
Accuracy		0.4	1.1	8673	1.1	0.99
Precision		0.4	1.9	8673	1.1	0.99
Intermediate Precision		0.8	2.0	8829	1.8	1.02
Solution Stability		0.3	2.2	8143	1.6	1.01
Robustness-I		0.2	1.7	8756	1.4	1.00
Robustness-II		0.4	1.4	9056	1.7	1.03

Robustness-III	0.3	1.0	9320	1.9	1.00
Robustness-IV	0.1	1.5	9810	1.3	0.98

Table-91: System Suitability for Pramipexole Imp E:

Experiment		% RSD		T.P	T.F.	Sim. Fac.
		RT	Area			
Specificity		0.5	1.7	12442	1.2	0.97
Linearity	LOQ	0.3	1.9	NA	NA	NA
	150%	0.1	0.4	NA	NA	NA
Accuracy		0.4	1.5	13211	1.0	0.99
Precision		0.4	1.5	13211	1.0	0.97
Intermediate Precision		0.5	2.4	11760	1.0	1.03
Solution Stability		0.4	2.0	10837	1.3	1.01
Robustness-I		0.6	1.9	12533	1.2	1.00
Robustness-II		0.4	1.2	11480	1.4	1.02
Robustness-III		0.5	1.5	12009	1.1	0.99
Robustness-IV		0.2	1.2	11689	1.3	1.01

Table-92: System Suitability for Pramipexole Imp D :

Experiment		% RSD		T.P	T.F.	Sim. Fac.
		RT	Area			
Specificity		0.5	1.1	12500	1.1	0.96
Linearity	LOQ	0.9	1.9	NA	NA	NA
	150%	0.2	0.2	NA	NA	NA
Accuracy		0.3	1.7	11783	1.1	0.96
Precision		0.3	1.7	11783	1.0	0.96
Intermediate Precision		0.6	2.9	12609	1.4	1.00

Solution Stability	0.3	2.5	11902	1.3	1.02
Robustness-I	0.2	2.5	10691	1.6	1.04
Robustness-II	0.5	1.2	11052	1.3	1.00
Robustness-III	0.7	1.8	13002	1.2	0.97
Robustness-IV	0.4	1.0	12604	1.0	1.02

The acceptance criterion for similarity factor was ≥0.95 ≤1.05, tailing factor for the peaks due to Pramipexole Imp A, Pramipexole Imp B, Pramipexole, Pramipexole Imp E and Pramipexole Imp D was NMT 5.0, theoretical plates was not less than 2000 and % RSD of RT , Area of Pramipexole Imp A, Pramipexole Imp B, Pramipexole, Pramipexole Imp E and Pramipexole Imp D Standard solution of first replicate solution was not more than 1.0% and 5.0% respectively. The acceptance criteria were met.

Blank (diluent) was injected to check the interference to Pramipexole Imp A, Pramipexole Imp B, Pramipexole, Pramipexole Imp E and Pramipexole Imp D. No interfernce to the peaks of Pramipexole Imp A, Pramipexole Imp B, Pramipexole, Pramipexole Imp E and Pramipexole Imp D.
The acceptance criteria were met.

4.2.2 LOD and LOQ:
A series of Standard Solutions of Pramipexole Imp A, Pramipexole Imp B, Pramipexole, Pramipexole Imp E and Pramipexole Imp D was prepared .and results are summarized as follows:

Table-93: Results of LOD and LOQ:

Components	LOD (ppm)	S/N Ratio	LOQ (ppm)	S/N Ratio
Pramipexole Imp A	0.054	4	0.75	13
Pramipexole Imp B	0.036	8	0.50	15
Pramipexole	0.032	5	0.50	12
Pramipexole Imp E	0.058	7	0.50	22
Pramipexole Imp D	0.095	6	0.50	16

4.2.3 Linearity:

Solutions containing concentrations of Pramipexole Imp A, Pramipexole Imp B, Pramipexole, Pramipexole Imp E and Pramipexole Imp D in the range of LOQ to 150% of the working level (i.e. 0.5 µg/ml to 7.5 µg/ml of Pramipexole Imp A, 0.3 µg/ml to 4.5 µg/ml of Pramipexole Imp B, 0.5 µg/ml to 7.5 µg/ml of Pramipexole, 0.5 µg/ml to 7.5 µg/ml of Pramipexole Imp D and 0.5 µg/ml to 7.5 µg/ml of Pramipexole Imp E) were injected into the chromatograph. The peak area responses were found to be linear with respect to the concentration.

The regression coefficient (r^2) for Pramipexole Imp A, Pramipexole Imp B, Pramipexole, Pramipexole Imp E and Pramipexole Imp D was determined as 0.9999, 0.9999, 0.9996 and 1.0000 respectively, the % Y-intercept for Pramipexole Imp A, Pramipexole Imp B, Pramipexole, Pramipexole Imp E and Pramipexole Imp D was 0.7%. -0.1%, 0.44% 0.0%, and 0.4% respectively, the

The criteria for regression coefficient was not less than 0.99, % Y-intercept was not more than ± 3.0%. The regression coefficient, the %Y-intercept, RSD of area and RT.

4.2.4 Precision -Repeatability:

The sample solution was prepared six times. The % of Imp calculation peak of Pramipexole Imp A, Pramipexole Imp B, Pramipexole Imp E and Pramipexole Imp D was not observed. Hence in accuracy and recovery chapter the 100% recovery level were injected in six times and calculated the % RSD of Pramipexole Imp A, Pramipexole Imp B, Pramipexole Imp E and Pramipexole Imp D recovery. The % RSD for the recovery of Pramipexole Imp A, Pramipexole Imp B, Pramipexole Imp E and Pramipexole Imp D of all six samples was 0.62%, 0.89%, 1.19 % and 0.94% respectively.

This met the acceptance criteria of not more than 5.00%.

4.2.5 Accuracy & Recovery:

The sample solution was prepared after spiking of known concentration in sample solution three times at each level (i.e. LOQ, 50%and 150%) and 100% level was injected in six replicates. The recovery of Pramipexole Imp A, Pramipexole Imp B, Pramipexole Imp E and

Pramipexole Imp D for each preparation was determined. The % RSD for the Pramipexole Imp A, Pramipexole Imp B, Pramipexole Imp E and Pramipexole Imp D recovery of samples was 0.68%, 0.88%, 0.74% and 0.57% for LOQ, 0.45%, 0.85%, 0.26% and 0.27% for 50% Level, 0.62%, 0.89%, 1.19 % and 0.94% for 100% level and 0.55% 0.72%, 1.18 and 0.96 % for 150% level respectively.

This met the acceptance criteria of not more than 5.00%.

4.2.6 Precision - Intermediate precision:

The sample solution was prepared six times. The % RSD for the recovery of Pramipexole Imp A, Pramipexole Imp B, Pramipexole Imp E and Pramipexole Imp D of all six samples was 1.18%, 0.82%, 1.29 % and 1.34% respectively.

The % variation of the % recovery of Pramipexole Imp A, Pramipexole Imp B, Pramipexole Imp E and Pramipexole Imp D from repeatability study and intermediate precision study was 1.01%, 1.01%, 1.09 % and 1.21% respectively. This met the acceptance criteria of not more than 5.0 %.

4.2.7 Solution Stability:

The % recovery of Pramipexole Imp A, Pramipexole Imp B, Pramipexole Imp E and Pramipexole Imp D was determined initially and then at the interval of 24 hrs. The % relative difference for the % Recovery of Pramipexole Imp A, Pramipexole Imp B, Pramipexole Imp E and Pramipexole Imp D was calculated at 24th hr time interval.

No additional peaks were observed for samples after the specified interval when compared with the initial run. Hence the sample solution was found to be stable upto 24 hours at laboratory conditions.

Table-94: % RSD for recovery of Impurities:

Validation Test	Observation	Result	Acceptance Criteria
Solution stability study	% RSD for recovery of Pramipexole Imp A			
	24 hrs	1.35 %	NMT 5.0%
	% RSD for recovery of Pramipexole Imp B			

	24 hrs	1.30 %	NMT 5.0%
	% RSD for recovery of Pramipexole Imp E		
	24 hrs	1.27 %	NMT 5.0%
	% RSD for recovery of Pramipexole Imp D		
	24 hrs	1.13%	NMT 5.0%

4.2.8 Robustness Exp-I:

Robustness Exp I was carried out by changing the flow rate of MP by – 0.1ml. The flow rate of MP changed from 0.8 ml to 0.7 ml per minute. Specificity, System suitability and the Precision of Samples were performed.

The sample solution was prepared three times. The % RSD for the recovery of Pramipexole Imp A, Pramipexole Imp B, Pramipexole Imp E and Pramipexole Imp D for each preparation was determined. The % RSD for the Pramipexole Imp A, Pramipexole Imp B, Pramipexole Imp E and Pramipexole Imp D content of all three samples was 1.04%, 0.96%, 0.72 % and 0.26% respectively.

The % Relative difference of the average content of Pramipexole Imp A, Pramipexole Imp B, Pramipexole Imp E and Pramipexole Imp D in precision-repeatability study and Robustness Exp. I study was 1.16%, 0.81%, 1.19 % and 1.09% respectively. This met the acceptance criteria of not more than 5.0%.

4.2.9 Robustness Exp-II:

Robustness Exp II was carried out by changing the flow rate of MP by +0.1ml. The flow rate of MP changed from 0.8 ml to 0.9 ml per minute. Specificity, System suitability and the Precision of Samples were performed.

The sample solution was prepared three times. The % RSD for the recovery of Pramipexole Imp A, Pramipexole Imp B, Pramipexole Imp E and Pramipexole Imp D for each preparation was determined. The % RSD for the Pramipexole Imp A, Pramipexole Imp B, Pramipexole Imp E and Pramipexole Imp D content of all three samples was 1.01%, 0.87%, 1.25 % and 1.16% respectively.

The % Relative difference of the average content of Pramipexole Imp A, Pramipexole Imp B, Pramipexole Imp E and Pramipexole Imp D in precision-repeatability study and Robustness Exp.

II study was 1.07%, 0.90%, 1.23 % and 1.22 % respectively. This met the acceptance criteria of not more than 5.0%.

4.2.10 Robustness Exp-III:

Robustness Exp-III was carried out by changing the column oven temperature by -2 °C. The column oven temperature changed from 50 °C to 48 °C. Specificity, System suitability and the Precision of Samples were performed.

The sample solution was prepared three times. The % RSD for the recovery of Pramipexole Imp A, Pramipexole Imp B, Pramipexole Imp E and Pramipexole Imp D for each preparation was determined. The % RSD for the Pramipexole Imp A, Pramipexole Imp B, Pramipexole Imp E and Pramipexole Imp D content of all three samples was 0.69%, 1.55%, 0.80 % and 1.33 % respectively.

The % Relative difference of the average content of Pramipexole Imp A, Pramipexole Imp B, Pramipexole Imp E and Pramipexole Imp D in precision-repeatability study and Robustness Exp.III study was 1.29%, 1.03%, 1.13 % and 1.25 % respectively. This met the acceptance criteria of not more than 5.0%.

4.2.11 Robustness Exp-IV:

Robustness Exp-IV was carried out by changing the column oven temperature by +2 °C. The column oven temperature changed from 50 °C to 52 °C. Specificity, System suitability and the Precision of Samples were performed.

The sample solution was prepared three times. The % RSD for the recovery of Pramipexole Imp A, Pramipexole Imp B, Pramipexole Imp E and Pramipexole Imp D for each preparation was determined. The % RSD for the Pramipexole Imp A, Pramipexole Imp B, Pramipexole Imp E and Pramipexole Imp D content of all three samples was 0.90%, 0.65%, 0.94 % and 0.89% respectively.

The % Relative difference of the average content of Pramipexole Imp A, Pramipexole Imp B, Pramipexole Imp E and Pramipexole Imp D in precision-repeatability study and Robustness Exp. IV study was 1.18%, 1.02%, 1.19 % and 1.17 % respectively. This met the acceptance criteria of not more than 5.0%.

4.3 Conclusion:

Based upon the data and results obtained for the analytical method validation, the HPLC analytical method for the determination of Assay and Related substances in Pramipexole tablets, is specific, selective and stability indicating. The method is linear in the range of 100.0 µg/ml to 300.0 µg/ml for Pramipexole of the specified limit. The method is precise, reproducible, rugged, accurate and robust to the changes in the analytical conditions. The HPLC analytical method for the determination of Assay and Related substances in Pramipexole Tablets is validated and can be accepted for the intended application.

The proposed HPLC method can be regarded as selective, accurate, precise, and valid for determination of acyclovir with a total running time of 10.0 min. Through this method it was possible to evaluate, Pramipexole quantification and offers advantages over methods previously reported. The present method is more sensitive, while the analytical run is shorter, permitting a high throughput.

The method of Assay and Related substances of Pramipexole tablets complied with the acceptance criteria set for the analytical parameters: Specificity & System suitability, Limit of detection and Limit of quantitation, Linearity, Precision-Repeatability, Accuracy and Recovery, Precision-Intermediate precision, Stability of sample and standard solution and Robustness studies. The method used for the Assay and Related substances of Pramipexole tablets complied with the acceptance criteria. Hence it is also used for routine analysis.

References:

1. Caroll D.I., Dzidic I, Stillwell R.N, Haegele K.D, Horning E.C.; *Analytical Chemistry*, 47, 2369- 2373, **(1975)**.
2. Karas M, Hillenkamp F; Laser desorption ionization of proteins with molecular masses exceeding 10,000 daltons, *Analytical Chemistry*, 60 ,2299-2301,**(1988)** .
3. Settle F.A; Handbook of instrumental techniques for analytical chemistry, ed; Pearson Education, Singapore, **(2004)**.
4. Khopkar S.M; .Basic Concept of Analytical Chemistry .Published by Wiley Eastern limited, New Delhi, **(1984)**.
5. Ahuja S, Scypinski S. (2001), Handbook of Modern Pharmaceutical Analysis. 1 st edition, volume 3,Academic Press, USA, 90-91.
6. Colombo P., Betini R., Peracchia M.T., Santi P.;Controlled Release Dosage Forms: From Ground to Space, *European Journal of Drug Metabolism and Pharmacokinetics*, 21, 87-91,**(1996)**.
7. Indian Pharmacopoiea (2010), 6 th edition, Indian Pharmacopoeia commission, Ghaziabad, **1**, 656-658.
8. USP 32 – NF 27, General Chapter 1225, Validation of Compendial Methods, 2009
9. USP 32 – NF 27, General Chapter 1226, Verification of Compendial Methods, 2009.
10. ICH Q3A(R)(2000), International Conferences on Harmonization, Draft Revised Guidance on Impurities in New Drug Substances. Federal Register, **65(140)**, 45085-45090.
11. ICH Q3B(R)(2000), International Conferences on Harmonization, Draft Revised Guidance on Impurities in New Drug Products. Federal Register, **65(139)**, 44791- 44797.
12. ICH Harmonized Tripartite Guideline, ICH Q2A, Text on Validation of Analytical procedures, June **(1995)**.
13. ICH Harmonized Tripartite Guideline, ICH Q2B, Validation of Analytical procedures: Methodology, June **(1997)**.
14. Reviewer Guidance: Validation of chromatographic Methods, Centre for Drug and Research .U.S. Government Printing office, Washington DC, **(1994)**.
15. Mierau J. (1995), Pramipexole: A dopamine-receptor agonist for treatment of Parkinson's Disease. *Clin Neuropharmacol.*, **18(Suppl 1)** , S195–S206.

16. Bouthenet ML, Souil E, Martres MP, Sokoloff P, Giros B, Schwartz JC (1991) , Localization of dopamine D3 receptor mRNA in the rat brain using in situ hybridization histochemistry: comparison with dopamine D2 receptor mRNA. *Brain Res* , **564(2)**,203-219.
17. Bennett JP Jr, Piercey MF, (1999), Pramipexole-a new dopamine agonist for the treatment of Parkinson's disease. *J Neurol Sci.* **163(1)**, 25-31.
18. Haselbarth VF, Justus-Obenauer H, Peil H, (1994), Pharmacokinectics and bioavailability of pramipexole: Comparison of plasma levels after intravenous and oral administration in healthy volunteers (M/2730/0029), *Upjohn Technical Report.*, **7215**, 94–96.
19. Wright CE, Sisson TL, Ichhpurani AK, Peters GR (1997), Steady-state pharmacokinetic properties of pramipexole in healthy volunteers., *J Clin Pharmacol.* **37(6)**, 520-525.
20. Wynalda MA, Wienkers LC (1997), Assessment of potential interactions between dopamine receptor agonists and various human cytochrome P450 enzymes using a simple in vitro inhibition screen. *Drug Metab Dispos.* **25 (10)**, 1211-1214.
21. Parkinson Study Group. (1997) , Safety and efficacy of pramipexole in early Parkinson disease. A randomized dose-ranging study. *JAMA.* , **278 (2)**, 125-130.
22. Shannon KM, Bennett JP Jr, Friedman JH (1997) , Efficacy of pramipexole, a novel dopamine agonist, as monotherapy in mild to moderate Parkinson's disease. *The Pramipexole Study Group. Neurology.* , **49(3)** ,724-728.
23. Bennett JP Jr, Piercey MF (1999) , Pramipexole--a new dopamine agonist for the treatment of Parkinson's disease. *J Neurol Sci.* ,**163(1)** ,25-31.
24. Parkinson Study Group. (2000) , Pramipexole vs levodopa as initial treatment for Parkinson disease: A randomized controlled trial. *JAMA.***284** ,1931–1938.
25. Lang AF. Munsat TL, (1999) , Quantification of Neurologic Deficit. Boston, Mass: Butterworth-Heinemann, 285–309.
26. Holloway RG, Shoulson I, Fahn S, (2004), Pramipexole vs levodopa as initial treatment for Parkinson disease a 4-year randomized controlled trial. *Arch Neurol.* **61** , 1044–1053.
27. Lieberman A, Ranhosky A, Korts D (1997) , Clinical evaluation of pramipexole in advanced Parkinson's disease: results of a double-blind, placebo-controlled, parallel-group study. *Neurology.* **49(1)** ,162-168.

28. Weiner WJ, Factor SA, Jankovic J, Hauser RA, Tetrud JW, Waters CH, Shulman LM, Glassman PM, Beck B, Paume D, Doyle C (2001), The long-term safety and efficacy of pramipexole in advanced Parkinson's disease. *Parkinsonism Relat Disord.*, **7(2)**, 115-120.
29. Möller JC, Oertel WH, Köster J, Pezzoli G, Provinciali L (2005), Long-term efficacy and safety of pramipexole in advanced Parkinson's disease: results from a European multicenter trial. *Mov Disord.*, **20(5)**, 602-610.
30. Wong KS, Lu CS, Shan DE, Yang CC, Tsoi TH, Mok V J (2003), Efficacy, safety, and tolerability of pramipexole in untreated and levodopa-treated patients with Parkinson's disease. *Neurol Sci.* **216(1)**, 81-87.
31. Parkinson Study Group (2007), Pramipexole in levodopa-treated Parkinson disease patients of African, Asian, and Hispanic heritage. *Clin Neuropharmacol.* **30(2)**, 72-85.
32. Poewe WH, Rascol O, Quinn N, Tolosa E, Oertel WH, Martignoni E, Rupp M, Boroojerdi B, (2007), Efficacy of pramipexole and transdermal rotigotine in advanced Parkinson's disease: a double-blind, double-dummy, randomised controlled trial. *SP 515 Investigators Lancet Neurol.* **6(6)**, 513-520.
33. Mizuno Y, Yanagisawa N, Kuno S, Yamamoto M, Hasegawa K, Origasa H, Kowa H, (2003), Randomized, double-blind study of pramipexole with placebo and bromocriptine in advanced Parkinson's disease. *Japanese Pramipexole Study Group Mov Disord.* **18(10)**, 1149-1156.
34. Guttman M (1997), Double-blind comparison of pramipexole and bromocriptine treatment with placebo in advanced Parkinson's disease. *International Pramipexole-Bromocriptine Study Group. Neurology.*, 1060-1065.
35. Rektorova I, Rektor I, Bares M, Dostal V, Ehler E, Fiedler J, Ressner P, (2003), Pramipexole and pergolide in the treatment of depression in Parkinson's disease: a national multicentre prospective randomized study. *Eur J Neurol.* **10(4)**, 399-406.
36. Hanna PA, Ratkos L, Ondo WG, Jankovic J (2001), Switching from pergolide to pramipexole in patients with Parkinson's disease. *J Neural Transm.* **108(1)**, 63-70.
37. Goetz CG, Blasucci L, Stebbins GT (1999), Switching dopamine agonists in advanced Parkinson's disease: is rapid titration preferable to slow?, *Neurology.* **52(6)**, 1227-1229.

38. Reichmann H, Odin P, Brecht HM, Koster J, Kraus PH (2006), Changing dopamine agonist treatment in Parkinson's disease: experiences with switching to pramipexole. *J Neural Transm Suppl.* **(71)** ,17-25.
39. Inzelberg R, Carasso RL, Schechtman E, Nisipeanu P (2000), Review A comparison of dopamine agonists and catechol-O-methyltransferase inhibitors in Parkinson's disease. *Clin Neuropharmacol.* **23(5)**, 262-266.
40. Navan P, Findley LJ, Jeffs JA, Pearce RK, Bain PG (2003), Randomized, double-blind, 3-month parallel study of the effects of pramipexole, pergolide, and placebo on Parkinsonian tremor. *Mov Disord.* **18(11)**, 1324-1331.
41. Pogarell O, Gasser T, van Hilten JJ, Spieker S, Pollentier S, Meier D, Oertel WH (2002), Pramipexole in patients with Parkinson's disease and marked drug resistant tremor: a randomised, double blind, placebo controlled multicentre study. *J Neurol Neurosurg Psychiatry.* **72(6)** , 713-720.
42. Lemke MR, Brecht HM, Koester J, Reichmann H (2006), Effects of the dopamine agonist pramipexole on depression, anhedonia and motor functioning in Parkinson's disease. *J Neurol Sci.* **248(1-2)**, 266-270.
43. Aarsland D, Zaccai J, Brayne C (2005), Review A systematic review of prevalence studies of dementia in Parkinson's disease. *Mov Disord.* **20(10)** ,1255-1263.
44. RektorovA I, Rektor I, Bares M, Dostál V, Ehler E, (2005) Cognitive performance in people with Parkinson's disease and mild or moderate depression: effects of dopamine agonists in an add-on to L-dopa therapy. *Eur J Neurol.* **12(1)**, 9-15.
45. Relja M, Klepac N (2006) , A dopamine agonist, pramipexole, and cognitive functions in Parkinson's disease. *J Neurol Sci.* **248(1-2)** , 251-254.
46. Biglan KM, Holloway RG Jr, McDermott MP, Richard IH, (2007), Risk factors for somnolence, edema, and hallucinations in early Parkinson disease. *CALM-PD Investigators Neurology.* **69(2)**, 187-195
47. Pluck GC, Brown RG. (2002), Apathy in Parkinson's disease. *J Neurol Neurosurg Psychiatry.* **73** ,636–642.
48. Guttman M, Jaskolka J (2001), The use of pramipexole in Parkinson's disease: are its actions D(3) mediated?, *Parkinsonism Relat Disord.* **7(3)**, 231-234.

49. Kumru H, Santamaria J, Valldeoriola F, Marti MJ, Tolosa E (2006), Increase in body weight after pramipexole treatment in Parkinson's disease. *Mov Disord.* **21(11)**, 1972- 1974.
50. Fantini ML, Gagnon JF, Filipini D, Montplaisir J (2003), The effects of pramipexole in REM sleep behavior disorder. *Neurology.* **61(10)**, 1418-1420.
51. Schmidt MH, Koshal VB, Schmidt HS (2006), Use of pramipexole in REM sleep behavior disorder: results from a case series. *Sleep Med.* **7(5)**, 418-423.
52. Schifitto GF, Oakes D, Shulman LM, (2006), levodopa-naive subjects with early Parkinson's disease, abstract at the First World Parkinson Congress. Washington DC, USA.
53. Holman AJ, Myers RR (2005), A randomized, double-blind, placebo-controlled trial of pramipexole, a dopamine agonist, in patients with fibromyalgia receiving concomitant medications. *Arthritis Rheum.* **52(8)**, 2495-2505.
54. Pogarell O, Gasser T, van Hilten JJ, Spieker S, Pollentier S, Meier D, Oertel WH (2002) ,Pramipexole in patients with Parkinson's disease and marked drug resistant tremor: a randomised, double blind, placebo controlled multicentre study. *J Neurol Neurosurg Psychiatry.* **72(6)**, 713-720.
55. Ahlskog JE, Muenter MD (2001), Frequency of levodopa-related dyskinesias and motor fluctuations as estimated from the cumulative literature. *Mov Disord.* **16(3)**, 448-458.
56. Parkinson Study Group (2000), Pramipexole vs levodopa as initial treatment for Parkinson disease: A randomized controlled trial. *JAMA.* **284(15)**, 1931-1938.
57. Holloway RG, Shoulson I, Fahn S, Kieburtz K, Lang A, Marek K, McDermott M, Seibyl J, Weiner W, Musch B, Kamp C, Welsh M, Shinaman A, Pahwa R, Barclay L, Hubble J, LeWitt P, Miyasaki J, Suchowersky O, (2004), Pramipexole vs levodopa as initial treatment for Parkinson disease: a 4-year randomized controlled trial. *Arch Neurol.* **61(7)**, 1044-1053.
58. Hauser RA, McDermott MP, Messing S (2006), Factors associated with the development of motor fluctuations and dyskinesias in Parkinson disease. *Arch Neurol.* **63(12)**, 1756-1760.
59. Inzelberg R, Schechtman E, Nisipeanu P (2003), Review Cabergoline, pramipexole and ropinirole used as monotherapy in early Parkinson's disease: an evidence-based comparison. *Drugs Aging.* **20(11)**, 847-855.

60. Izumi Y, Sawada H, Yamamoto N, Kume T, Katsuki H, Shimohama S, Akaike A (2007), Novel neuroprotective mechanisms of pramipexole, an anti-Parkinson drug, against endogenous dopamine-mediated excitotoxicity. *Eur J Pharmacol.* **57(2-3)**, 132-140.
61. Presgraves SP, Borwege S, Millan MJ, Joyce JN (2004), Involvement of dopamine D(2)/D(3) receptors and BDNF in the neuroprotective effects of S32504 and pramipexole against 1-methyl-4-phenylpyridinium in terminally differentiated SH-SY5Y cells. *Exp Neurol.* **190(1)**, 157-170.
62. Gu M, Iravani MM, Cooper JM, King D, Jenner P, Schapira AH (2004), Pramipexole protects against apoptotic cell death by non-dopaminergic mechanisms. *J Neurochem.* **91(5)**, 1075-1081.
63. Uberti D, Bianchi I, Olivari L, Ferrari-Toninelli G, Canonico P, Memo M (2007), Pramipexole prevents neurotoxicity induced by oligomers of beta-amyloid. *Eur J Pharmacol.* **569(3)**, 194-196.
64. Carvey PM, McGuire SO, Ling ZD (2001), Neuroprotective effects of D3 dopamine receptor agonists. *Parkinsonism Relat Disord.* **7(3)**, 213-223.
65. Sayeed I, Parvez S, Winkler-Stuck K, Seitz G, Trieu I, Wallesch CW, Schönfeld P, Siemen D (2006), Patch clamp reveals powerful blockade of the mitochondrial permeability transition pore by the D2-receptor agonist pramipexole. *FASEB J.* **20(3)**, 556-558.
66. Joyce JN, Woolsey C, Ryoo H, Borwege S, Hagner D (2004), Low dose pramipexole is neuroprotective in the MPTP mouse model of Parkinson's disease, and downregulates the dopamine transporter via the D3 receptor. *BMC Biol.* **11**, 20-22.
67. Parkinson Study Group (2002), Dopamine transporter brain imaging to assess the effects of pramipexole vs levodopa on Parkinson disease progression. *JAMA.* **287(13)**, 1653-1661.
68. Ravina B, Eidelberg D, Ahlskog JE, Albin RL, Brooks DJ, Carbon M, Dhawan V, Feigin A, Fahn S, Guttman M, Gwinn-Hardy K, McFarland H, Innis R, Katz RG, Kieburtz K, Kish SJ, Lange N, Langston JW, Marek K, Morin L, Moy C, Murphy D, Oertel WH, Oliver G, Palesch Y, Powers W, Seibyl J, Sethi KD, Shults CW, Sheehy P, Stoessl AJ, Holloway R (2005), Review The role of radiotracer imaging in Parkinson disease. *Neurology.* **64(2)**, 208-215.
69. Guttman M, Jaskolka J (2001), The use of pramipexole in Parkinson's disease: are its actions D(3) mediated?, *Parkinsonism Relat Disord.* **7(3)**, 231-234.

70. Constantinescu R, Romer M, McDermott MP, Kamp C, Kieburtz K (2007), Impact of pramipexole on the onset of levodopa-related dyskinesias. CALM-PD Investigators of the Parkinson Study Group, *Mov Disord.* **22(9)**, 1317-1319.
71. Noyes K, Dick AW, Holloway RG (2004), Pramipexole v. levodopa as initial treatment for Parkinson's disease: a randomized clinical-economic trial. Parkinson Study Group *Med Decis Making.* **24(5)**, 472-485.
72. Noyes K, Dick AW, Holloway RG (2005), Pramipexole and levodopa in early Parkinson's disease: dynamic changes in cost effectiveness., Parkinson Study Group *Pharmacoeconomics.* **23(12)**, 1257-1270.
73. Moller JC, Oertel WH, Koster J, Pezzoli G, Provinciali L (2005), Long-term efficacy and safety of pramipexole in advanced Parkinson's disease: results from a European multicenter trial. *Mov Disord.* **20(5)**, 602-610.
74. Paus S, Brecht HM, Köster J, Seeger G, Klockgether T, Wüllner U (2003), Sleep attacks, daytime sleepiness, and dopamine agonists in Parkinson's disease. *Mov Disord.* **18(6)**, 659-667.
75. Holloway RG Jr, McDermott MP, Richard IH, (2007), Risk factors for somnolence, edema, and hallucinations in early Parkinson disease. *Parkinson Study Group CALM-PD Investigators Neurology.* **69(2)**, 187-195.
76. Etminan M, Samii A, Takkouche B, Rochon PA (2001), Increased risk of somnolence with the new dopamine agonists in patients with Parkinson's disease: a meta-analysis of randomised controlled trials. *Drug Saf.* **24(11)**, 863-868.
77. Etminan M, Gill S, Samii A (2003), Review Comparison of the risk of adverse events with pramipexole and ropinirole in patients with Parkinson's disease: a meta-analysis. *Drug Saf.* **26(6)**, 439-444.
78. Happe S, Berger K, FAQT Study Investigators (2001), The association of dopamine agonists with daytime sleepiness, sleep problems and quality of life in patients with Parkinson's disease--a prospective study. *J Neurol.* **248(12)**, 1062-1967.
79. O'Suilleabhain PE, Dewey RB Jr (2002), Contributions of dopaminergic drugs and disease severity to daytime sleepiness in Parkinson disease. *Arch Neurol.* **59(6)**, 986-989.

80. Hobson DE, Lang AE, Martin WR, Razmy A, Rivest J, Fleming J (2002), Excessive daytime sleepiness and sudden-onset sleep in Parkinson disease: a survey by the Canadian Movement Disorders Group. *JAMA*. **287(4)**, 455-463.
81. Homann CN, Wenzel K, Suppan K, Ivanic G, Kriechbaum N, Crevenna R, Ott E (2005), Review Sleep attacks in patients taking dopamine agonists. *BMJ*. **324(7352)**, 1483-1487.
82. Avorn J, Schneeweiss S, Sudarsky LR, Benner J, Kiyota Y, Levin R, Glynn RJ (2005), Sudden uncontrollable somnolence and medication use in Parkinson disease. *Arch Neurol*. **62(8)**, 1242-1248.
83. Romigi A, Brusa L, Marciani MG, Pierantozzi M, Placidi F, Izzi F, Sperli F, Testa F, Stanzione P (2005), Sleep episodes and daytime somnolence as result of individual susceptibility to different dopaminergic drugs in a PD patient: a polysomnographic study. *J Neurol Sci*. **228(1)**, 7-10.
84. Razmy A, Lang AE, Shapiro CM (2004), Predictors of impaired daytime sleep and wakefulness in patients with Parkinson disease treated with older (ergot) vs newer (nonergot) dopamine agonists. *Arch Neurol*. **61(1)**, 97-102.
85. Nieves AV, Lang AE (2002), Treatment of excessive daytime sleepiness in patients with Parkinson's disease with modafinil. *Clin Neuropharmacol*. **25(2)**, 111-114.
86. Adler CH, Caviness JN, Hentz JG, Lind M, Tiede J (2003), Randomized trial of modafinil for treating subjective daytime sleepiness in patients with Parkinson's disease. *Mov Disord*. **18(3)**, 287-293.
87. Ondo WG, Fayle R, Atassi F, Jankovic J (2005), Modafinil for daytime somnolence in Parkinson's disease: double blind, placebo controlled parallel trial. *J Neurol Neurosurg Psychiatry*. **76(12)**, 1636-1639.
88. Peralta C, Wolf E, Alber H, Seppi K, Müller S, Bösch S, Wenning GK, Pachinger O, Poewe W (2006), Valvular heart disease in Parkinson's disease vs. controls: An echocardiographic study. *Mov Disord*. **21(8)**, 1109-1113.
89. Zanettini R, Antonini A, Gatto G, Gentile R, Tesei S, Pezzoli G (2007), Valvular heart disease and the use of dopamine agonists for Parkinson's disease. *N Engl J Med*. **356(1)**, 39-46.
90. Junghanns S, Fuhrmann JT, Simonis G, Oelwein C, Koch R, Strasser RH, Reichmann H, Storch A (2007), Valvular heart disease in Parkinson's disease patients treated with

dopamine agonists: a reader-blinded monocenter echocardiography study. *Mov Disord.* **22(2)**, 234-238.

91. Schade R, Andersohn F, Suissa S, Haverkamp W, Garbe E (2007), Dopamine agonists and the risk of cardiac-valve regurgitation. *N Engl J Med.* **356(1)**, 29-38.
92. Dewey RB 2nd, Reimold SC, O'Suilleabhain PE (2007), Cardiac valve regurgitation with pergolide compared with nonergot agonists in Parkinson disease. *Arch Neurol.* **64(3)**, 377-380.
93. Rothman RB, Baumann MH, Savage JE, Rauser L, McBride A, Hufeisen SJ, Roth BL (2000), Evidence for possible involvement of 5-HT(2B) receptors in the cardiac valvulopathy associated with fenfluramine and other serotonergic medications. *Circulation.* **102(23)**, 2836-2841.
94. Chaudhuri KR, Dhawan V, Basu S, Jackson G, Odin P (2004), Valvular heart disease and fibrotic reactions may be related to ergot dopamine agonists, but non-ergot agonists may also not be spared. *Mov Disord.* **19(12)**, 1522-1523.
95. Dodd ML, Klos KJ, Bower JH, Geda YE, Josephs KA, Ahlskog JE (2005), Pathological gambling caused by drugs used to treat Parkinson disease. *Arch Neurol.* **62(9)**, 1377-1381.
96. Klos KJ, Bower JH, Josephs KA, Matsumoto JY, Ahlskog JE (2005), Pathological hypersexuality predominantly linked to adjuvant dopamine agonist therapy in Parkinson's disease and multiple system atrophy. *Parkinsonism Relat Disord.* **11(6)**, 381-386.
97. Weintraub D, Siderowf AD, Potenza MN, Goveas J, Morales KH, Duda JE, Moberg PJ, Stern MB (2006), Association of dopamine agonist use with impulse control disorders in Parkinson disease. *Arch Neurol.* **63(7)**, 969-973.
98. Mamikonyan E, Siderowf AD, Duda JE, Potenza MN, Horn S, Stern MB, Weintraub D (2008), Long-term follow-up of impulse control disorders in Parkinson's disease. *Mov Disord.* **23(1)**, 75-80.
99. Quickfall J, Suchowersky O (2007), Pathological gambling associated with dopamine agonist use in restless legs syndrome. *Parkinsonism Relat Disord.* **13(8)**, 535-536.
100. Lu C, Bharmal A, Suchowersky O (2006), Gambling and Parkinson disease. *Arch Neurol.* **63(2)**,298.
101. Parkinson Study Group (2000), Pramipexole vs levodopa as initial treatment for Parkinson disease: A randomized controlled trial. *JAMA.* **284(15)**, 1931-1938.

102. Mizuno Y, Yanagisawa N, Kuno S, Yamamoto M, Hasegawa K, Origasa H, Kowa H, (2003), Randomized, double-blind study of pramipexole with placebo and bromocriptine in advanced Parkinson's disease. *Japanese Pramipexole Study Group Mov Disord.* **18(10)**, 1149-1156.
103. Biglan KM, Holloway RG Jr, McDermott MP, Richard IH (2007), Risk factors for somnolence, edema, and hallucinations in early Parkinson disease. *Neurology.* **69(2)**, 187-195.
104. Tan EK, Ondo W (2000), Clinical characteristics of pramipexole-induced peripheral edema. *Arch Neurol.* **57(5)**, 729-732.
105. Kleiner-Fisman G, Fisman DN (2007), Risk factors for the development of pedal edema in patients using pramipexole. *Arch Neurol.* **64(6)**, 820-824.
106. IUPAC Technical Report, Harmonized Guidelines for Single-Laboratory Validation of Methods of Analysis, Pure Appl. Chem., **74 (5)**, 835-855, 2002.
107. Eurachem – The Fitness for Purpose of Analytical Methods A Laboratory Guide to Method Validation and Related Topics, 1998
108. AOAC, How to Meet ISO 17025 Requirements for Methods Verification, 2007
109. L. Huber, Validation and Qualification in Analytical Laboratories, Informa Healthcare, New York, USA, 2007
110. C. T. Viswanathan Workshop/Conference Report Quantitative Bio-analytical Methods Validation and Implementation: Best Practices for Chromatographic and Ligand Binding Assays. AAPS Journal, **9(1)**, E30-E42, 2007.
111. ICH Q2B, Validation of Analytical Procedures: Methodology, adopted in **1996**, Geneva Q2B, in 2005 incorporated in Q2(R1)
112. ICH Q2A, Validation of Analytical Procedures: Definitions and Terminology, Geneva, 1995, in 2005 incorporated in Q2(R1)
113. Roy J (2002), Pharmaceutical Impurities- A Mini-Review, *AAPS Pharm.Sci.Tech.* **3(2)**, 1-8.
114. Ahuja S (2007), Assuring quality of drugs by monitoring impurities, *Adv. Drug Deliv.* 3-11.
115. Wishweshwar S and Gupta R.M.(2002), FAQ's on Imp Profile for Bulk Drugs, *Pharma Times*, **34(3)**, 13-15

116. Roy J, Pharmaceutical Impurities- A Mini-Review, AAPS Pharm.Sci.Tech. 3(2), 6, 2002, 1-8.
117. United State Pharmacopeia (USP26/NF21) United State Pharmacopeial Convention, Inc. Rockville: 2003, 2331.
118. ICH Q3A(R), International Conferences on Harmonization, Draft Revised Guidance on Impurities in New Drug Substances. Federal Register, 65(140), 2000, 45085-45090.
119. Roy J, (2002) Pharmaceutical Impurities- A Mini-Review, AAPS Pharm.Sci.Tech. 3(2), 6, 1-8
120. McGovern T. and David j. (2006), Regulation of genotoxic and carcinogenic impurities in drug substances and products, *Trends Anal. Chem.* **25(8)**, 790-795.
121. ICH Q3A(R), International Conferences on Harmonization, Draft Revised Guidance on Impurities in New Drug Substances. Federal Register, 65(140), 2000,
122. ICH Q3B(R), International Conferences on Harmonization, Draft Revised Guidance on Impurities in New Drug Products. Federal Register, 65(139)
123. Indian Pharmacopoiea, 6 th edition, Indian Pharmacopoeia commission, Ghaziabad, Volume I, 2010, 656-658.
124. Ahuja S (2007), Assuring quality of drugs by monitoring impurities, *Adv. Drug Deliv.*, 3-11.
125. Patel K.N. (2010), Introduction to hyphenated techniques and their applications in pharmacy, *Pharm. Methods.* **1(1)**, 2-13.
126. Ahuja S, Scypinski S. (2001), Handbook of Modern Pharmaceutical Analysis. 1 st edition, volume 3, Academic Press, USA, 90-91.
127. Oliver MD (2002), Imp Profiling of Pholcodeine by Liquid Chromatography Electrospray Ionization Mass spectrometry (LC-ESI-MS), *J. Pharmacy and Pharmacol.***54**, 87-96.
128. David C. (2002), The Role of LC-MS in Drug Discovery, *Recent Application in LC-MS*, **11, 2-6**.
129. Ermer J. (1998), a quality concept for impurities during drug development – use of hyphenated LC – MS technique, *Pharm. Sci.Technol.*, **1(2)**, 76 – 82.
130. Patel KN, (2010) Introduction to hyphenated techniques and their applications in harmacy, Pharm. Methods. Oct-Dec, 2010, 1(1), 2-13.

131. ICH Q3A(R)(2000), International Conferences on Harmonization, Draft Revised Guidance on Impurities in New Drug Substances. Federal Register, **65(140)**, 45085-45090.
132. ICH Q3B(R)(2000), International Conferences on Harmonization, Draft Revised Guidance on Impurities in New Drug Products. Federal Register, **65(139)**, 44791- 44797.
133. ICH Harmonized Tripartite Guideline, ICH Q2A, Text on Validation of Analytical procedures, June **(1995)**.

CPSIA information can be obtained
at www.ICGtesting.com
Printed in the USA
LVHW031129070423
743748LV00002B/242